Mathematical Tricks, Puzzles and Oddities for the Young and Old

Problem Solving in Mathematics and Beyond

Print ISSN: 2591-7234
Online ISSN: 2591-7242

Series Editor: Dr. Alfred S. Posamentier
Distinguished Lecturer
New York City College of Technology - City University of New York

---

There are countless applications that would be considered problem solving in mathematics and beyond. One could even argue that most of mathematics in one way or another involves solving problems. However, this series is intended to be of interest to the general audience with the sole purpose of demonstrating the power and beauty of mathematics through clever problem-solving experiences.

Each of the books will be aimed at the general audience, which implies that the writing level will be such that it will not be engulfed in technical language — rather the language will be simple everyday language so that the focus can remain on the content and not be distracted by unnecessarily sophiscated language. Again, the primary purpose of this series is to approach the topic of mathematics problem-solving in a most appealing and attractive way in order to win more of the general public to appreciate this most important subject rather than to fear it. At the same time we expect that professionals in the scientific community will also find these books attractive, as they will provide many entertaining surprises for the unsuspecting reader.

Published

Vol. 47 *Mathematical Tricks, Puzzles and Oddities for the Young and Old*
by Alfred S Posamentier

Vol. 46 *Surprising Geometric Solutions: A Projective Geometry Approach*
by Alfred S Posamentier and David Stuhlpfarrer

Vol. 45 *How to Generate a Genuine Love of Mathematics*
by Alfred S Posamentier

Vol. 44 *Revolutions in Mathematics*
by Arthur D Kramer and Alfred S Posamentier

Vol. 43 *Mathematical Puzzles and Curiosities*
by Ivo Fagundes David de Oliveira, Tanya Khovanova and Yogev Shpilman

For the complete list of volumes in this series, please visit www.worldscientific.com/series/psmb

Problem Solving in
Mathematics and Beyond
Volume 47

Mathematical Tricks, Puzzles and Oddities for the Young and Old

Alfred S. Posamentier
The City University of New York, USA

World Scientific

NEW JERSEY • LONDON • SINGAPORE • BEIJING • SHANGHAI • HONG KONG • TAIPEI • CHENNAI • TOKYO

Published by

World Scientific Publishing Co Pte Ltd
5 Toh Tuck Link, Singapore 596224
USA office: 27 Warren Street, Suite 401-402, Hackensack, NJ 07601
UK office: 57 Shelton Street, Covent Garden, London WC2H 9HE

Library of Congress Cataloging-in-Publication Data
Names: Posamentier, Alfred S. author
Title: Mathematical tricks, puzzles and oddities for the young and old /
Alfred S. Posamentier, The City University of New York, USA.
Description: New Jersey : World Scientific, [2027] | Series: Problem solving in
mathematics and beyond, 2591-7234 ; Vol. 47 | Includes index.
Identifiers: LCCN 2026018879 | ISBN 9789819824328 hardcover |
ISBN 9789819823185 paperback | ISBN 9789819823192 ebook |
ISBN 9789819823208 ebook other
Subjects: LCSH: Mathematical recreations | Problem solving--Mathematics
Classification: LCC QA95 .P659 2027
LC record available at https://lccn.loc.gov/2026018879

British Library Cataloguing-in-Publication Data
A catalogue record for this book is available from the British Library.

For any available supplementary material, please visit
https://www.worldscientific.com/worldscibooks/10.1142/14583#t=suppl

Desk Editors: Kannan Krishnan/Gabriel Rawlinson

Typeset by Stallion Press
Email: enquiries@stallionpress.com

About the Author

Alfred S. Posamentier was most recently a Distinguished Lecturer at New York City College of Technology of the City University of New York. Prior to that he was Executive Director for Internationalization and Funded Programs at Long Island University, New York. This was preceded by 5 years as Dean of the School of Education and Professor of Mathematics Education at Mercy University, New York. Before that he was for 40 years at the City College of the City University of New York, at which he is now Professor Emeritus of Mathematics Education and Dean Emeritus of the School of Education. He is the author and co-author of more than 90 mathematics books for teachers, secondary and elementary school students, as well as the general readership. Dr. Posamentier is also a frequent commentator in newspapers and journals on topics related to education and mathematics.

After completing his B.A. degree in mathematics at Hunter College of the City University of New York, he took a position as a teacher of mathematics at Theodore Roosevelt High School (Bronx, New York), where he focused his attention on improving the students' problem-solving skills and at the same time enriching their instruction far beyond what the traditional textbooks offered. During his six-year tenure there, he also developed the school's first mathematics teams (both at the junior and senior level). He is still involved in working with mathematics teachers and supervisors, nationally and internationally, to help them maximize their effectiveness.

Immediately upon joining the faculty of the City College of New York in 1970 (after having received his master's degree there in 1966), he began to develop in-service courses for secondary school mathematics teachers, including such special areas as recreational mathematics and problem solving in mathematics. As Dean of the City College School of Education for 10 years, his scope of interest covered the full gamut of educational issues. During his tenure as Dean, he took the School of Education from the bottom of the New York State rankings to the top with a perfect NCATE accreditation assessment in 2009. He also raised more than $12,000,000 from the private sector for innovative educational programs. Posamentier repeated this successful transition at Mercy University, where he enabled it to become the only college to have received both NCATE and TEAC accreditation simultaneously.

In 1973, Dr. Posamentier received his Ph.D. from Fordham University (New York) in mathematics education, and he has since extended his reputation in mathematics education to Europe. He has been visiting professor at several European universities in Austria, England, Germany, the Czech Republic, Turkey and Poland. In 1990, he was Fulbright Professor at the University of Vienna.

In 1989 he was awarded an Honorary Fellow position at the South Bank University (London, England). In recognition of his outstanding teaching, the City College Alumni Association named him Educator of the Year in 1994, and in 2009. New York City had the day, May 1, 1994, named in his honor by the President of the New York City Council. In 1994, he was also awarded the *Grosse Ehrenzeichen für Verdienste um die Republik Österreich* (Grand Medal of Honor from the Republic of Austria), and in 1999, upon approval of Parliament, the President of the Republic of Austria awarded him the title of University Professor of Austria. In 2003, he was awarded the title of *Ehrenbürgerschaft* (Honorary Fellow) of the Vienna University of Technology, and in 2004 he was awarded the *Österreichisches Ehrenkreuz für Wissenschaft & Kunst 1. Klasse* (Austrian Cross of Honor for Arts and Science,

First Class) from the President of the Republic of Austria. In 2005, he was inducted into the Hunter College Alumni Hall of Fame, and in 2006 he was awarded the prestigious Townsend Harris Medal by the City College Alumni Association. He was inducted into the New York State Mathematics Educator's Hall of Fame in 2009, and in 2010 he was awarded the coveted Christian-Peter-Beuth Prize from the Technische Fachhochschule—Berlin. In 2017, Posamentier was awarded *summa cum laude nemine discrepante* by the Fundacion Sebastian, A.C., Mexico City, Mexico.

He has taken on numerous important leadership positions in mathematics education locally. He was a member of the New York State Education Commissioner's Blue Ribbon Panel on the Math-A Regents Exams, and the Commissioner's Mathematics Standards Committee, which redefined the Mathematics Standards for New York State, and he also served on the New York City Schools' Chancellor's Math Advisory Panel.

Dr. Posamentier is still a leading commentator on educational issues and continues his long-time passion of seeking ways to make mathematics interesting to teachers, students and the general public—as can be seen from some of his more recent books.

For more information and a list of his publications see: https://en.wikipedia.org/wiki/Alfred_S._Posamentier.

Contents

Introduction

Many people enjoy clever oddities. Challenging mathematical oddities can be thought provoking and rewarding to solve. One might even be motivated to share these amazing and surprising oddities with others. These oddities could be simple exercises with an unusual and unexpected mathematical result. Some of these unexpected results can be justified and explained. Other amazing results have escaped explanation for centuries. In any case, both can be entertaining and stimulating for the brain. This latter aspect should not be taken for granted because exercising the brain keeps it sharp and allows it to grow. This is perhaps in sync with a popular proverb: "He/She who rests, rusts." We offer here a variety of activities that will keep the brain active, and the reader entertained.

There are times when the concept of the prank has a mischievous tone. Rest assured that the oddities are all in a favorable ambience. The oddities in this book are aimed at the general audience, so non-mathematicians need not fear. The oddities will cover a wide variety of subjects in elementary mathematics. They will flex your numerical, arithmetical, geometrical, and even your logical abilities. You can challenge yourself, and thereby enhance your mental capabilities and self-confidence. Not only will they be stimulating, but they will also entertain and provide new insights into mathematics.

We welcome you now to begin your journey through a plethora of entertaining and highly unusual aspects of elementary mathematics. Not only should you enjoy the surprising results, but we hope that you will share these amazements with friends and family.

Chapter 1

The Magic of Numbers

This chapter explores the extraordinary aspects of numbers, beginning with a creative look at the number 1 and continuing through unusual traits of individual numbers, revealing surprising relationships between numbers and types of numbers. The reader will be entertained and amazed at our number system's hidden secrets.

The Magic of Numbers Consisting of Only 1s

Numbers consisting of only 1s are often called *repunits*, which stems from "repeated units." Such numbers can be generated by the following formula:

$$r_n = \frac{10^n - 1}{9}.$$

The number 11 is the smallest repunit, and is a prime number. The next largest number after 11 that consists of all 1s is the number 111, and it, too, has some curious properties. For example, the differences of two squares can be selected equal to a number consisting of all 1s.

The progression of such differences of squares leading to numbers consisting of repunit numbers follows:

$$1^2 - 0^2 = 1$$
$$6^2 - 5^2 = 11$$
$$20^2 - 17^2 = 111$$
$$56^2 - 45^2 = 1{,}111$$
$$156^2 - 115^2 = 11{,}111$$
$$556^2 - 445^2 = 111{,}111$$
$$344^2 - 85^2 = 111{,}111$$
$$356^2 - 125^2 = 111{,}111.$$

Within this list of numbers that will result in 1s, we can see another pattern emerging if we look at the second, fourth, and sixth entries. Each time, an additional 5 and 4 is tagged onto the front of each term. Continue this pattern and notice what a spectacular pattern evolves.

$$6^2 - 5^2 = 11$$
$$56^2 - 45^2 = 1{,}111$$
$$556^2 - 445^2 = 111{,}111$$
$$5556^2 - 4445^2 = 11{,}111{,}111$$
$$55556^2 - 44445^2 = 1{,}111{,}111{,}111$$
$$555556^2 - 444445^2 = 111{,}111{,}111{,}111$$
$$5555556^2 - 4444445^2 = 11{,}111{,}111{,}111{,}111$$
$$55555556^2 - 44444445^2 = 1{,}111{,}111{,}111{,}111{,}111$$
$$555555556^2 - 444444445^2 = 111{,}111{,}111{,}111{,}111{,}111$$

...

$$55555555555555556^2 - 44444444444444445^2 =$$
$$1{,}111{,}111{,}111{,}111{,}111{,}111{,}111{,}111{,}111{,}111{,}111.$$

Of this list, the only prime number is 11. In fact, the next two prime numbers composed of all 1s are 1,111,111,111,111,111,111, and 11,111,111,111,111,111,111,111. These two numbers will be prime in any arrangement of the digits—since all the digits are the same!

Then there are more curious numbers, such as those that are generated by the difference of squares and are multiples of numbers consisting of digits that are all 1s.

$$7^2 - 4^2 = 33 = 3 \times 11$$
$$67^2 - 34^2 = 3{,}333 = 3 \times 1{,}111$$
$$667^2 - 334^2 = 333{,}333 = 3 \times 111{,}111$$
$$6667^2 - 3334^2 = 33{,}333{,}333 = 3 \times 11{,}111{,}111$$
$$66667^2 - 33334^2 = 3{,}333{,}333{,}333 = 3 \times 1{,}111{,}111{,}111.$$

Here is another such pattern of numbers, which should be admired.

$$8^2 - 3^2 = 55 = 5 \times 11$$
$$78^2 - 23^2 = 5555 = 5 \times 1{,}111$$
$$778^2 - 223^2 = 555{,}555 = 5 \times 111{,}111$$
$$7778^2 - 2223^2 = 55{,}555{,}555 = 5 \times 11{,}111{,}111$$
$$77778^2 - 22223^2 = 5{,}555{,}555{,}555 = 5 \times 1{,}111{,}111{,}111.$$

Our investigation of repunit numbers brings us to an interesting pattern. We divide 111,111,111 by 9, which gives us the number 12,345,679. Notice that the digits are in numerical order, but we are missing the number 8. Yet, when we consider the pattern shown in Figure 1-1, the 8 is once again included in a pattern generating numbers consisting of only 1s.

0	×	9	+	1	=	1
1	×	9	+	2	=	11
12	×	9	+	3	=	111
123	×	9	+	4	=	1,111
1,234	×	9	+	5	=	11,111
12,345	×	9	+	6	=	111,111
123,456	×	9	+	7	=	1,111,111
1,234,567	×	9	+	8	=	11,111,111
12,345,678	×	9	+	9	=	111,111,111
123,456,789	×	9	+	10	=	1,111,111,111

Figure 1-1

As you can see, repunit numbers can generate some rather interesting relationships and patterns. Another curious pattern evolves when we take the square of successive unit-digit numbers, as shown in Figure 1-2.

Number of 1s	n	n^2
1	1	1
2	11	121
3	111	12321
4	1111	1234321
5	11111	123454321
6	111111	12345654321
7	1111111	1234567654321
8	11111111	123456787654321
9	111111111	12345678987654321
10	1111111111	1234567900987654321

Figure 1-2

To get a better view of these repunit numbers, r_n, we will factor them into their prime factors, as shown in Figure 1-3.

We notice that r_2 and r_{19} are prime numbers. The question then arises, are there other such prime repunits? The answer is yes. However, mathematicians have struggled with this question for years. For example, the German mathematician Carl Gustav Jacob Jacobi (1804–1851) pursued the question as to whether the repunit r_{11} is a prime number. Today, a computer system can answer this question in less than a second. Factoring repunit numbers is often very difficult, however with the aid of a computer we find that the repunit r_{71} is factorable as follows, and, therefore, not a prime number:

$$\begin{aligned} R_{71} &= 11{,}111{,}111{,}111{,}111{,}111{,}111{,}111{,}111{,}111{,}111{,}111{,}111{,}111{,} \\ &\quad 111{,}111{,}111{,}111{,}111{,}111{,}111{,}111{,}111{,}111 \\ &= 241{,}573{,}142{,}393{,}627{,}673{,}576{,}957{,}439{,}049 \times \\ &\quad 45{,}994{,}811{,}347{,}886{,}846{,}310{,}221{,}728{,}895{,}223{,}034{,}301{,}839. \end{aligned}$$

R_1	=	1	=	1
r_2	=	**11**	=	**11**
r_3	=	111	=	3×37
r_4	=	1,111	=	11×101
r_5	=	11,111	=	41×271
r_6	=	111,111	=	3×7×11×13×37
r_7	=	1,111,111	=	239×4,649
r_8	=	11,111,111	=	11×73×101×137
r_9	=	111,111,111	=	3^2 ×37×333,667
r_{10}	=	1,111,111,111	=	11×41×271×9,091
r_{11}	=	11,111,111,111	=	21,649×513,239
r_{12}	=	111,111,111,111	=	3×7×11×13×37×101×9,901
r_{13}	=	1,111,111,111,111	=	53×79×265,371,653
r_{14}	=	11,111,111,111,111	=	11×239×4,649×909,091
r_{15}	=	111,111,111,111,111	=	3×31×37×41×271×2,906,161
r_{16}	=	1,111,111,111,111,111	=	11×17×73×101×137×5,882,353
r_{17}	=	11,111,111,111,111,111	=	2,071,723×5,363,222,357
r_{18}	=	111,111,111,111,111,111	=	3^2×7×11×13×19×37×52,579×333,667
r_{19}	=	**1,111,111,111,111,111,111**	=	**1,111,111,111,111,111,111**
r_{20}	=	11,111,111,111,111,111,111	=	11×41×101×271×3,541×9,091×27,961

Figure 1-3

By 1930, it was known that r_2 and r_{19} (Oscar Hoppe, 1916) as well as r_{23} (Lehmer and Kraitchik, 1929) are prime numbers.[1] In 1970, a mathematics student, E. Seah, was able to demonstrate that the repunit r_{317} is also a prime number. The search for a prime number among the repunits continues. For example, in 1985 the repunit r_{1031} was discovered to be a prime by H. C. Williams and H. Dubner.

[1] Richard L. Francis. "Mathematical Haystacks: Another Look at Repunit Numbers." *The College Mathematics Journal*, Vol. 19, No. 3. (May, 1988), pp. 240–246.

Further repunits that have been since identified as primes are: r_{49081} (H. Dubner, 1999), r_{86453} (L. Baxter, 2000), r_{109297} (P. Bourdelais, H. Dubner, 2007), and r_{270343} (M. Voznyy, A. Budnyy, 2007).[2] It is believed today that there are an infinite number of repunits which are prime numbers.

Fabulous Number Patterns

Here are some noteworthy and curious patterns. Consider the equation $192 + 384 = 576$. At first glance you might not notice that every other digit from left to right contains the sequence 1, 2, 3, 4, 5, and 6, while 7, 8, and 9 are placed in the center positions from right to left.

Another amazing symmetry appears when you take the difference of the numbers 123,456,789 and 987,654,321. That is, $987{,}654{,}321 - 123{,}456{,}789 = 864{,}197{,}532$, which is a number that contains each of the nine digits exactly once.

Another curiosity can be found in the calculation $291{,}548{,}736 = 8 \times 92 \times 531 \times 746$. Again, we notice that on either side of the equation each of the nine digits are used exactly once.

We can take this a step further to show two products, each of which uses all nine digits exactly once: $124{,}367{,}958 = 627 \times 198{,}354 = 9 \times 26 \times 531{,}487$. On each side of the equal signs all nine digits appear exactly once.

There is also a number squared which delivers each of the digits exactly twice: $335{,}180{,}136^2 = 112{,}345{,}723{,}568{,}978{,}496$. There are also squares that produce a product where each of the numbers involved uses the nine digits exactly once, such as $493{,}827{,}156^2 = 246{,}913{,}578 \times 987{,}654{,}312$.

While on the topic of square roots, here are two numbers that are reversals of each other, where the square root of each is an integer: $\sqrt{10036224} = 3168$, and $\sqrt{42263001} = 6501$.

While we are dealing with the numbers from 1 to 9, we should notice that the smallest perfect square using all the digits is $\sqrt{139854276} = 11826$, and the largest perfect square using all nine digits is $\sqrt{923187456} = 30384$. If we were to include the zero, the smallest

[2] https://mathworld.wolfram.com/Repunit.html.

perfect square using all 10 digits is $\sqrt{1026753849} = 32043$, and the largest perfect square using all 10 digits is $\sqrt{9814072356} = 99066$.

To revisit the repunits we have an amazing relationship, which is $123{,}456{,}789 + 987{,}654{,}321 + 1 = 1{,}111{,}111{,}111$.

While we are considering all nine digits, we present an expression with fractions using all nine digits, resulting in the number 1, as follows:

$$\frac{5}{34}+\frac{7}{68}+\frac{9}{12}=\frac{30}{204}+\frac{21}{204}+\frac{153}{204}=1.$$

An Inexplicable Occurrence of the Number 1

Here is a "magical" property in mathematics that has baffled mathematicians for many years. To this day, no one knows why it happens. It is known as the *Collatz conjecture*, named after the German mathematician Lothar Collatz (1910–1990). The process begins with any number but always ends with the number 1. There are two rules that have to be followed as you work with any *arbitrarily* selected number.

- *If a given number is odd, then multiply it by 3 and add 1.*
- *If the number is even, then divide the number by 2.*

Continue this process until you reach the number 1. Regardless of the number originally selected, the result will *always* be 1.

Let's try it for the *arbitrarily selected* number **13.**
Since 13 is odd, we multiply by 3 and add 1 to get 40.
Since 40 is even, we simply divide by 2 to get 20.
Since 20 is also even so we divide by 2 to get 10.
The number 10 is even, so we divide by 2 to get 5.
Since 5 is odd, we multiply by 3 and add 1 to get 16.
Since 16 is even, we divide by 2 to get 8.
Since 8 is even, we divide by 2 to get 4.
Since 4 is even, we divide by 2 to get 2.
Since 2 is even, we divide by 2 to get 1, which is what the Collatz conjecture predicted.

Regardless of which number we begin with, we will eventually get to 1. This is truly remarkable! Had we started with 17 as our arbitrarily-selected starting number, we would have required 13 steps to reach the number 1. Starting with the number 43 will require 27 steps to reach the number 1. And amazingly, beginning with the number 97 will require 118 steps to reach the number 1.

The starting value having the most steps less than:

10 is 9, which has 19 steps;
100 is 97, which has 118 steps;
1000 is 871, which has 178 steps;
10,000 is 6,171, which has 261 steps;
100,000 is 7,7031, which has 350 steps;
1,000,000 is 837,799, which has 524 steps;
10,000,000 is 8,400,511, which has 685 steps;
100,000,000 is 63,728,127, which has 949 steps;
1,000,000,000 is 670,617,279, which has 986 steps.

Does this really work for all numbers? This question has concerned mathematicians since the 1930s, and to date no answer has been found, despite monetary rewards having been offered for a proof of this conjecture. Using computers, this problem, known in the literature as the "$3n + 1$ Problem," has been shown to be true for all the numbers up to $10^{18} - 1$. The longest progression of steps, so far established, is for the number 931,386,509,544,713,451, which requires 2283 steps to reach the number 1. So, you can see how the number 1 is omnipresent.

The Inescapable Number 1

Unlike the Collatz conjecture shown in the previous discussion, there are other techniques that generate the number 1. Here's one that we can justify using simple algebra. This goes as follows:

- Select a random number
- Double it and add 2 to the result
- Take half the number
- Subtract the number you began with

The result will always be 1.

Let's try this with the number 18. We double 18 to get 36, whereupon we add 2 to get 38. Then we take half, which is 19, and then subtract the original number, 18, to get 19 – 18 = 1. As opposed to the Collatz conjecture, which enabled us to reach the number 1 in a rather mysterious way, this procedure can be easily justified algebraically as follows:

$$(2x+2)\cdot\frac{1}{2}-x=x+1-x=1.$$

This shows you the power of algebra. Unfortunately, to the present day, no one has been able to construct a proof that the Collatz conjecture holds true for all numbers.

Curious Representation of the Number 1

With the digits 1, 2, 4, 5, 6, 7, 8, and 9, the challenge is to represent the number 1 using these digits exactly once.

There are clever solutions, of which we offer two here:

$$1=1^{23456789}$$
$$1=1+(2-3)+(5-4)+(6-7)+(9-8).$$

Special Prime Numbers

There are some curious prime numbers where all arrangements of their digits result in another prime number. A simple example would be to take the number 79. Each of the digits is a prime number, and the other arrangement of digits, 97, is also prime number. The first few prime numbers where all digits arrangements are prime numbers are 11, 13, 17, 37, 79, 113, 199, and 337. You might like to find the next few such primes that determine new prime numbers with all arrangements of their digits.

Symmetric Digits

This is merely for amusement. Notice each line in Figure 1-4 has the same digits in the same order on either side of the = sign, with the exponents in a lead position.

$42 \div 3 = 4 \times 3 + 2$	$\sqrt{121} = 12 - 1$
$63 \div 3 = 6 \times 3 + 3$	$\sqrt{64} = 6 + \sqrt{4}$
$95 \div 5 = 9 + 5 + 5$	$\sqrt{49} = 4 + \sqrt{9} = 9 - \sqrt{4}$
$(2+7) \times 2 \times 16 = 272 \div 16$	$\sqrt{169} = 16 - \sqrt{9} = \sqrt{16} + 9$
$5^{6-2} = 625$	$\sqrt{256} = 2 \times 5 + 6$
$(8+9)^2 = 289$	$\sqrt{324} = 3 \times (2 + 4)$
$2^{10} - 2 = 1022$	$\sqrt{11881} = 118 - 8 - 1$
$4 \times 2^3 = 4^3 \div 2 = 34 - 2$	$\sqrt{1936} = -1 + 9 + 36$
	$\sqrt[3]{1331} = 1 + 3 + 3 + 1 + 3$

Figure 1-4

Unusual Number Patterns

Numbers lend themselves to various patterns whose discovery is most enchanting. For example, we can partition the natural number sequence where we begin with an even number of numbers on the left side of the equality and an odd number of numbers on the right side. The next equality continues the sequentially, with an odd number of numbers on the left side and an even number of numbers on the right side. This can continue indefinitely.

$$1+2=3=3\left(1^2\right)$$
$$4+5+6=7+8=3\left(1^2+2^2\right)$$
$$9+10+11+12=13+14+15=3\left(1^2+2^2+3^2\right)$$
$$16+17+18+19+20=21+22+23+24=3\left(1^2+2^2+3^2+4^2\right)$$
$$25+26+27+28+29+30=31+32+33+34+35=3\left(1^2+2^2+3^2+4^2+5^2\right).$$

The Pythagorean theorem can also create some number patterns. The most common of the Pythagorean relationships is the equation $3^2+4^2=5^2$.

Now we find other consecutive numbers whose squares are equal following a pattern analogous to that above.

$$10^2+11^2+12^2=13^2+14^2$$
$$21^2+22^2+23^2+24^2=25^2+26^2+27^2$$
$$36^2+37^2+38^2+39^2+40^2=41^2+42^2+43^2+44^2.$$

A motivated reader may seek other such relationships where an even and odd number of numbers can be found on opposite sides of the equality.

Select Sum of Cubes

Figure 1-5 presents numbers that are multiplied by the sum of their digits to equal the sum of the cubes of the digits. Remarkable!

$$37 \times (3+7) = 3^3 + 7^3$$
$$48 \times (4+8) = 4^3 + 7^3$$
$$111 \times (11+1) = 11^3 + 1^3$$
$$147 \times (14+7) = 14^3 + 7^3$$
$$148 \times (14+8) = 14^3 + 8^3$$

Figure 1-5

Beautiful Growth of Numbers

There is not much that can be said about pattern shown in Figure 1-6,

$$16 = 4^2$$
$$1156 = 34^2$$
$$111156 = 334^2$$
$$11115556 = 3334^2$$
$$111111555556 = 333334^2$$
$$9 = 3^2$$
$$1089 = 33^2$$
$$110889 = 333^2$$
$$11108889 = 3333^2$$
$$1111088889 = 33333^2$$
$$111110888889 = 333333^2$$
$$11111108888889 = 3333333^2$$
$$1111111088888889 = 33333333^2$$
$$111111110888888889 = 333333333^2$$
$$11111111108888888889 = 3333333333^2$$
$$36 = 6^2$$
$$4356 = 66^2$$
$$443556 = 666^2$$
$$44435556 = 6666^2$$
$$4444355556 = 66666^2$$
$$444443555556 = 666666^2$$
$$44444435555556 = 6666666^2$$
$$4444444355555556 = 66666666^2$$
$$444444443555555556 = 666666666^2$$

Figure 1-6

Equal Reversals

An amazing phenomenon is shown in Figure 1-7, where each pair of numbers has the same product when the factors have the digits reversed.

$$12 \times 42 = 21 \times 24$$
$$12 \times 63 = 21 \times 36$$
$$12 \times 84 = 21 \times 48$$
$$13 \times 62 = 31 \times 26$$
$$23 \times 96 = 32 \times 69$$
$$24 \times 63 = 42 \times 36$$
$$24 \times 84 = 42 \times 48$$
$$26 \times 93 = 62 \times 39$$
$$36 \times 84 = 63 \times 48$$
$$46 \times 96 = 64 \times 69$$

More on digits reversals as follows:

$$42^2 + 53^2 + 68^2 = 24^2 + 85^2 + 36^2$$
$$42^2 + 58^2 + 63^2 = 24^2 + 85^2 + 36^2$$
$$43^2 + 52^2 + 68^2 = 34^2 + 25^2 + 86^2$$
$$43^2 + 58^2 + 62^2 = 34^2 + 85^2 + 26^2$$
$$48^2 + 52^2 + 63^2 = 84^2 + 25^2 + 36^2$$
$$48^2 + 53^2 + 62^2 = 84^2 + 35^2 + 26^2$$

$$13 + 42 + 53 + 57 + 68 + 97 = 79 + 86 + 75 + 35 + 24 + 31$$

$$13^2 + 42^2 + 53^2 + 57^2 + 68^2 + 97^2 = 79^2 + 86^2 + 75^2 + 35^2 + 24^2 + 31^2$$

$$13^3 + 42^3 + 53^3 + 57^3 + 68^3 + 97^3 = 79^3 + 86^3 + 75^3 + 35^3 + 2^3 + 31^3$$

Figure 1-7

Odd–Even Products and Sums

A most unusual relationship is shown in Figure 1-8. Notice how two numbers have a product and sum that are digit reversals and of different parities: one is odd, and the other, even.

$9 + 9 = 18$	$9 \times 9 = 81$
$24 + 3 = 27$	$24 \times 3 = 72$
$47 + 2 = 49$	$47 \times 2 = 94$
$497 + 2 = 499$	$497 \times 2 = 994$

Figure 1-8

The Curiosities of the Number 6

Some numbers have a property that is rarely known. For example, when multiplying an even number by 6, the units digit will always be the same as the units digit of the even number you multiplied by 6. For example, suppose we multiply $548 \times 6 = 3288$. Notice that the factor 548 and the product 3288 have the same units digit, 8. To convince the reader further, consider $784 \times 6 = 4704$. Notice that the units digit of the number multiplied by 6 is the same as the product, namely, 4.

Now for the amazing relationships that the number 6 generates:

$$
\begin{aligned}
&66+6=12\times 6\\
&666+66+6=123\times 6\\
&6666+666+66+6=1234\times 6\\
&66666+6666+666+66+6=12{,}345\times 6\\
&666{,}666+66{,}666+6666+666+66+6=123{,}456\times 6\\
&6{,}666{,}666+666{,}666+66{,}666+6666+666+66+6\\
&\quad =1{,}234{,}567\times 6\\
&66{,}666{,}666+6{,}666{,}666+666{,}666+66{,}666+6666+666+66+6\\
&\quad =12{,}345{,}678\times 6\\
&666{,}666{,}666+66{,}666{,}666+6{,}666{,}666+666{,}666+66{,}666+6666+\\
&666+66+6=123{,}456{,}789\times 6.
\end{aligned}
$$

The digit 6 is clearly dominant in this amazing pattern.

Also consider these arrangements of the digit 6:

$$
\begin{aligned}
666+666&=(6\times 6\times 6\times 6)+(6\times 6)\\
666+666&=66\left(6+6+6+\frac{6}{6}\right)+66+6+6\\
666+666&=\left(\frac{66}{6}\right)^3+\frac{6}{6}.
\end{aligned}
$$

Also, $666 = 123 + 1^2 + 2^2 + 3^2 + (1 \times 23^2) = 132 + 23^2 + 3 + 2$. We notice that the number 666 has been represented in various forms using only the digits 1, 2, and 3.

The number 666 is also related to permutations of the numbers 1, 3, and 5, as we see when we consider the sum of all the permutations: $153 + 351 + 135 + 531 + 315 + 513 = 2 \times 666$. Furthermore, the permutations can be separated as an equality of the reflections, so that $53 + 351 + 531 = 135 + 531 + 351$. By the way, $153 = 51 \times 3$, and $1 \times 5 \times 3 = 1^0 + 5^1 + 3^2$.

We can also represent the number 666 using only the digits 1, 2, 3, 4, 5, 6, 7, 8, and 9, in that order: $666 = 1 + 2 + 3 + 4 + 567 + 89$, and $666 = 123 + 456 + 78 + 9$. And considering the sequence in reverse numerical order, we have $666 = 9 + 87 + 6 + 543 + 21$.

The Missing Number 7

Here is a challenge regarding the number 7. Consider the set of integers 1000, 1001, 1002, 1003, ..., 9998, 9999. How many of these numbers do not contain the digit 7?

It would be tedious to list all the numbers in the sequence and then select those that do not contain the digit 7. The clever strategy would be to inspect the four-digit numbers and consider how many possible digits each place can have. For the units digit, the tens digit, and the hundreds digit, excluding 7 means each place can have one of nine different digits available. The thousandth place only has eight possible digits, since it cannot include a 0 or a 7. Therefore, to count the number of numbers that satisfy our requirement, we simply multiply $8 \times 9 \times 9 \times 9 = 5{,}832$.

Interesting Relationship Between the Numbers 8 and 11

Relationships between the numbers 8 and 11 are rare and strange, as they are not the same parity—one is a power of 2 and the other is a prime number. Yet, if we raise the numbers 8 and 11 to the same even power, the difference of these quantities will always be a multiple of 57. This is a rather amazing discovery. To convince you that this strange relationship exists, take the difference of the following even powers of 8 and 11:

$11^2 - 8^2 = 121 - 64 = 57$
$11^4 - 8^4 = 14{,}641 - 4{,}096 = 10{,}545 = 57 \times 185$
$11^6 - 8^6 = 1{,}771{,}561 - 262{,}144 = 1{,}509{,}417 = 57 \times 26{,}481$
$11^8 - 8^8 = 214{,}358{,}881 - 16{,}777{,}216 = 197{,}581{,}665 = 57 \times 3{,}466{,}345$
$11^{10} - 8^{10} = 25937424601 - 1073741824 = 24863682777 = 57 \times 436204961.$

Some Curiosities Involving the Number 9

There are many curiosities involving the number 9, largely because it is 1 less than the base 10 of the number system. One such curiosity is that the sum of the first nine consecutive prime numbers is a perfect square: $2 + 3 + 5 + 7 + 11 + 13 + 17 + 19 + 23 = 100$. Another curiosity involving the number 9 multiplies the ninth prime number, 23, by 9 to yield 207, the sum of whose digits is 9. It is always fun to discover these unusual relationships.

When we consider prime numbers, we should also notice that the only two prime numbers whose sum and difference are also both prime are 2 and 5, since $2 + 5 = 7$ (a prime number) and $5 - 2 = 3$ (also a prime number). Furthermore, the units digit of prime numbers is always a prime number with one exception, and that is the number 9, which is the only nonprime units digit in the list of prime numbers shown in Figure 1-9. (It is just a coincidence that this is the ninth figure of this chapter!)

We also find that the number 9 can be expressed in fractional form using all 10 digits exactly once, as shown with the following fractions:

$$9 = \frac{95742}{10638}, \ 9 = \frac{95823}{10647}, \text{ and } 9 = \frac{97524}{10836}.$$

However, if we allow the 0 to take a first position in any of the numbers, we get an additional three fractional equivalents to the number 9, namely,

$$9 = \frac{57429}{06381}, \ 9 = \frac{58239}{06471}, \text{ and } 9 = \frac{75249}{08361}.$$

Prime Numbers Less Than 1000												
2	3	5	7	11	13	17	19	23	29	31	37	41
43	47	53	59	61	67	71	73	79	83	89	97	101
103	107	109	113	127	131	137	139	149	151	157	163	167
173	179	181	191	193	197	199	211	223	227	229	233	239
241	251	257	263	269	271	277	281	283	293	307	311	313
317	331	337	347	349	353	359	367	373	379	383	389	397
401	409	419	421	431	433	439	443	449	457	461	463	467
479	487	491	499	503	509	521	523	541	547	557	563	569
571	577	587	593	599	601	607	613	617	619	631	641	643
647	653	659	661	673	677	683	691	701	709	719	727	733
739	743	751	757	761	769	773	787	797	809	811	821	823
827	829	839	853	857	859	863	877	881	883	887	907	911
919	929	937	941	947	953	967	971	977	983	991	997	

Figure 1-9

Thus, we can see how the number 9, from its place of 1 below the number 10, has many unique characteristics. But there are more to follow.

The Inevitable Number 9

One of the most amazing relationships involving the number 9 can be found when you take any multidigit number, create another number with the same digits by scrambling them around into other arrangements, and then subtract these two numbers. The result will always be a multiple of 9. Perhaps this is best clarified by considering a random number, say, 385,627 and then randomly rearranging the digits to get 285,763. The difference of these two numbers is 99,864, which is equal to $9 \times 11{,}096$. To further understand this surprising relationship, consider one more example: $94{,}273 - 29{,}374 = 64{,}899 = 9 \times 7211$.

Strange Multiplications by 9

The number 9 provides us with some more amazing patterns, such as when we multiply $12{,}345{,}679 \times 9 = 111{,}111{,}111$. If we multiply the same number 12,345,679 by the number 9^2, we will arrive in another unexpected pattern, namely, $12{,}345{,}679 \times 81 = 999{,}999{,}999$. Suppose we multiply 9 by the reversal of this factor, this time including the number 8 in the digits. We arrive at another pattern provided by the number 9, as follows: $987{,}654{,}321 \times 9 = 8{,}888{,}888{,}889$. Analogous patterns can be seen in the following Figures 1-10 and 1-11.

$$
\begin{array}{rcccccl}
0 & \times & 9 & + & 1 & = & 1 \\
1 & \times & 9 & + & 2 & = & 11 \\
12 & \times & 9 & + & 3 & = & 111 \\
123 & \times & 9 & + & 4 & = & 1{,}111 \\
1{,}234 & \times & 9 & + & 5 & = & 11{,}111 \\
12{,}345 & \times & 9 & + & 6 & = & 111{,}111 \\
123{,}456 & \times & 9 & + & 7 & = & 1{,}111{,}111 \\
1{,}234{,}567 & \times & 9 & + & 8 & = & 11{,}111{,}111 \\
12{,}345{,}678 & \times & 9 & + & 9 & = & 111{,}111{,}111
\end{array}
$$

Figure 1-10

The number 9 also provides an opportunity for us to look back at the number 8, which appears repeatedly in Figure 1-11.

$$
\begin{array}{rcccccl}
0 & \times & 9 & + & 8 & = & 8 \\
9 & \times & 9 & + & 7 & = & 88 \\
98 & \times & 9 & + & 6 & = & 888 \\
987 & \times & 9 & + & 5 & = & 8{,}888 \\
9{,}876 & \times & 9 & + & 4 & = & 88{,}888 \\
98{,}765 & \times & 9 & + & 3 & = & 888{,}888 \\
987{,}654 & \times & 9 & + & 2 & = & 8{,}888{,}888 \\
9{,}876{,}543 & \times & 9 & + & 1 & = & 88{,}888{,}888 \\
98{,}765{,}432 & \times & 9 & + & 0 & = & 888{,}888{,}888
\end{array}
$$

Figure 1-11

This time you can use the 9 in another way to generate an amazing pattern, which is shown in Figure 1-12.

$$
\begin{array}{rclcl}
9 & \times & 9 & = & 81 \\
99 & \times & 99 & = & 9{,}801 \\
999 & \times & 999 & = & 998{,}001 \\
9{,}999 & \times & 9{,}999 & = & 99{,}980{,}001 \\
99{,}999 & \times & 99{,}999 & = & 9{,}999{,}800{,}001 \\
999{,}999 & \times & 999{,}999 & = & 999{,}998{,}000{,}001 \\
9{,}999{,}999 & \times & 9{,}999{,}999 & = & 99{,}999{,}980{,}000{,}001
\end{array}
$$

Figure 1-12

Justifying an Oddity of the Number 9

We sometimes overlook strange coincidences or oddities, and when we do see them, we try to justify them. Consider this very unusual relationship:

> *Any two-digit number ending in 9 can be expressed as the sum of the product of the digits and sum of the digits.*

More simply stated, *any two-digit number ending in 9 = [product of digits] + [sum of digits].*

When we are confronted with a mathematical oddity, we might use algebra to explain the unusual pattern. This is the case here.

Before we justify this strange relationship, we might want to see specific examples such as these:

$$\begin{aligned}
9 &= (0 \cdot 9) + (0 + 9)\\
19 &= (1 \cdot 9) + (1 + 9)\\
29 &= (2 \cdot 9) + (2 + 9)\\
39 &= (3 \cdot 9) + (3 + 9)\\
49 &= (4 \cdot 9) + (4 + 9)\\
59 &= (5 \cdot 9) + (5 + 9)\\
69 &= (6 \cdot 9) + (6 + 9)\\
79 &= (7 \cdot 9) + (7 + 9)\\
89 &= (8 \cdot 9) + (8 + 9)\\
99 &= (9 \cdot 9) + (9 + 9).
\end{aligned}$$

One of algebra's advantages is that it allows us to justify many mathematical curiosities. Do not be too enchanted with a nice pattern. We are using it as a means to an end, and not an end in itself. That is, we would like to use this as an illustration of algebra explaining an oddity in mathematics.

We can represent a two-digit number as $10t + u$, where t represents the tens digit and u represents the units digit. Then the sum of the digits is $t + u$ and the product of the digits is tu.

The number meeting the above conditions $= 10t + u = (tu) + (t + u)$. Then $10t = tu + t$, when $u = 9$, and $t \neq 0$.

This discussion should evoke a curiosity about numbers with more than two digits. For example:

$$109 = (10 \cdot 9) + (10 + 9)$$
$$119 = (11 \cdot 9) + (11 + 9)$$
$$129 = (12 \cdot 9) + (12 + 9).$$

Here, the digits to the left of the 9 are treated as a discrete number t, just as we treated the tens digit previously. The results are the same. This can be extended to any number of digits as long as the units digit is 9. For example, consider the number 7849. Using the above scheme, we get $784 + 9 = 793$ and $784 \cdot 9 = 7056$. Amazingly, their sum is $793 + 7056 = 7849$, which is the number we started with. This is a technique which we were able to justify algebraically.

Tracing Sequential Numbers

While we are considering numbers with patterns to their digits, Figure 1-13 shows a pattern whose beauty is inexpressible through words.

11×11	=	121
111×111	=	12,321
1111 ×1111	=	1,234,321
11111×11111	=	123,454,321
111111 × 111111	=	12,345,654,321
1111111 × 1111111	=	1,234,567,654,321
11111111 × 11111111	=	123,456,787,654,321
111111111 × 111111111	=	12,345,678,987,654,321

Figure 1-13

An Oddity of the Number 11

Having discovered unusual aspects of the number 9, as it is the number 1 less than the base 10, we will now find a plethora of curiosities embedded in the number 11, as it is 1 greater than the base 10. One such curiosity takes any number which is a multiple of 11 and reversing the digits to generate another number which is also a multiple of 11. For example, the number 135,795 = 11 × 12,345. We reverse the digits of the original number to get 597,531 = 54,321 × 11. Notice that the number and the multiplier are both reversals of the original product.

A Number 11 Surprise

The number 11 lends itself to some more very curious mathematical surprises. Suppose you have a number composed of multiple digits where no two adjacent digits have a sum greater than 9. Then you can do the following:

- multiply the number by 11,
- reverse the digits of this product,
- divide the resulting number by 11,

and you will find that you have attained a number whose digits are the reverse of the original number. For example, consider 36,181,451 × 11 = 397,995,961. We now reverse the digits and divide that number by 11 to get: 169,599,793 ÷ 11 = 15,418,163, which unexpectedly is a reversal of the original number.

A Strange Creation for a Multiple of 11

Begin with a number which has an even number of digits, such that the first and last digit are either both even numbers or odd numbers. Then add to this number its reversal. The result will always be a multiple of 11, as we can see from the following example.

Start with the number 542,189. The number with the digits reversed is 981,245. Both numbers' initial and terminal digits are odd numbers. When we add these two numbers, 542,189 + 981,245 = 1,523,434 = 11 × 138,494. Naturally, this scheme can also work where the initial and terminal digits are even, such as 851,376 + 673,158 = 1,524,534 = 11 × 138,594. Remember, you must begin with an even number of digits, and the initial and terminal digits must be of the same parity.

An Oddity of the Number 11

Oddities with the amazing number 11 are offered to further amuse the reader further. First, we represent 11^2 as the sum of five consecutive powers of 3 as follows:

$$11^2 = 121 = 3^0 + 3^1 + 3^2 + 3^3 + 3^4.$$

Then we represent 11^3 as the sum of the squares of three consecutive odd numbers:

$$11^3 = 1331 = 19^2 + 21^2 + 23^2.$$

A motivated reader may seek other such oddities with the number 11.

Oddities of the Number 13

Besides being seen by some as an unlucky number, the number 13 does have some special features in mathematics:

- $13^2 = 7 + 8 + 9 + 10 + 11 + 12 + 13 + 14 + 15 + 16 + 17 + 18 + 19$.
- The sum of the first thirteen natural numbers $1 + 2 + 3 + 4 + 5 + 6 + 7 + 8 + 9 + 10 + 11 + 12 + 13 = 91$, which is the thirteenth triangular number.
- The sum of the first six prime numbers $2 + 3 + 5 + 7 + 11 + 13 = 41$, which is the thirteenth prime number.
- The sum of the first thirteen prime numbers $2 + 3 + 5 + 7 + 11 + 13 + 17 + 19 + 23 + 29 + 31 + 37 + 41 = 238$, which has a digit sum of $2 + 3 + 8 = 13$.
- There are exactly thirteen prime numbers consisting of four consecutive digits, and they are: 1423, 2143, 2341, 2543, 4231, 4253, 4523, 4567, 4657, 5647, 5867, 6547, and 6857.

- The number of composite numbers between 13 and its reversal, 31, is thirteen. These composite numbers are: 14, 15, 16, 18, 20, 21, 22, 24, 25, 26, 27, 28, 30.

Another Number 13 Curiosity

There are only thirteen palindromic numbers whose digit sum is 13. They are:

373, 12,721 14,341 31,513 32,323 1,117,111 1,215,121 1,411,141 3,211,123 112,212,211 113,111,311 121,131,121 131,111,131.

Notice how all of these numbers are palindromic numbers, and their digit sum is 13. Also notice that zeros were intentionally not included. The number 13 seems to possess nice features, despite its reputation as an unlucky number.

Coincidences with the Number 19

For the most part, the prime number 19 is pretty much neglected. However, when we search for curious appearances of 19, we notice for example that $19^{19} = 1{,}978{,}419{,}655{,}660{,}313{,}589{,}123{,}979$, where the first two digits are 19. Furthermore, 19 is the only prime number equal to the differences of two prime cubes, that is, $3^3 - 2^3 = 19$. There is another curious relationship involving the number 19: Its reversal is 91, and $91^2 = 8281$, whose digit sum $8 + 2 + 8 + 1 = 19$.

Another coincidence involving the number 19 and its reversal 91 is that the product of these two numbers is 7029, which is the smallest number that can be expressed as the sum of two cubes in two different ways, that is, $1729 = 1^3 + 12^3 = 9^3 + 10^3$.

The Strange Relationship Between the Numbers 19 and 9109

Some number relationships are discovered accidentally and without justification. It's also difficult sometimes to describe the relationship. For example, Figure 1-14 shows a relationship where words are not needed to describe these unusual patterns.

1 × 19 = 19	1 + 9 = 10	1 + 0 = 1	**1**
2 × 19 = 38	3 + 8 = 11	1 + 1 = 2	**2**
3 × 19 = 57	5 + 7 = 12	1 + 2 = 3	**3**
4 × 19 = 76	7 + 6 = 13	1 + 3 = 4	**4**
5 × 19 = 95	9 + 5 = 14	1 + 4 = 5	**5**
6 × 19 = 114	11 + 4 = 15	1 + 5 = 6	**6**
7 × 19 = 133	13 + 3 = 16	1 + 6 = 7	**7**
8 × 19 = 152	15 + 2 = 17	1 + 7 = 8	**8**
9 × 19 = 171	17 + 1 = 18	1 + 8 = 9	**9**
10 × 19 = 190	19 + 0 = 19	1 + 9 =10	**10**

Figure 1-14

In Figure 1-15, the number 9109 generates some amazing relationships when multiplied by the numbers 1–9. Inspecting the vertical list of products shows that the units digits are in descending numerical order, beginning with the number 9 and ending with the number 1. The tens digits begin with 0 and end with the number 8. The hundreds digits are simply 1–9, and the thousands digits are 9–1. Once again, the ten thousands progress in sequence. Furthermore, the sums of the digits of each of the numbers listed in the third column also present a surprising sequence of numbers: 19–27. All this stemming from the harmless number 9109.

	Product	Digit Sum
9109 × 1 =	0 9 1 0 9	19
9109 × 2 =	1 8 2 1 8	20
9109 × 3 =	2 7 3 2 7	21
9109 × 4 =	3 6 4 3 6	22
9109 × 5 =	4 5 5 4 5	23
9109 × 6 =	5 4 6 5 4	24
9109 × 7 =	6 3 7 6 3	25
9109 × 8 =	7 2 8 7 2	26
9109 × 9 =	8 1 9 8 1	27

Figure 1-15

The Magic of the Number 21

Unexpectedly, the number 21 has a very unusual property that is often neglected. When you select any multiple of 21, such as $21 \times 18 = 378$, delete the units digit to get 37, multiply this number $37 \times 11 = 407$, and then subtract the previously deleted units digit 8, you get 399, which amazingly is a multiple of 21, since $399 = 21 \times 19$.

To firm up an understanding of this procedure, we will consider another example. We begin by getting a multiple of 21, in this case, $21 \times 634 = 13{,}314$. The next step is to delete the units digit leaving us with the number 1331, which we then multiply by 11 to get 14,641, from which we then subtract the previously deleted units digit, 4, to get 14,637. Once again, amazingly we find that $14{,}637 = 21 \times 697$.

The Mysterious Number 22

We will consider a mathematics trick that can be easily justified algebraically. Begin by following the instructions very carefully.

- Select any three-digit number with all digits different from one another.
- Write all possible two-digit numbers that can be formed from the three digits of the selected number.
- Then divide the sum of these two-digit numbers by the sum of the digits in the original three-digit number.
- Quite unexpectedly, you will arrive at the number 22.

Let's consider as an example the three-digit number 365. Take the sum of all the possible two-digit numbers that can be formed from these three digits: $36 + 35 + 63 + 53 + 65 + 56 = 308$. The sum of the digits of the original number is $3 + 6 + 5 = 14$. We then divide these two numbers $\frac{308}{14} = 22$. This trick will work for any three-digit number as long as the three digits are not alike.

On occasion it is nice to justify a remarkable numerical phenomenon. Therefore, we will now analyze this unusual result beginning with a general representation of the number: $100x + 10y + z$. We now find the sum of all the two-digit numbers taken from the three digits:

$$\begin{aligned}(10x+y)&+(10y+x)+(10x+z)+(10z+x)+(10y+z)+(10z+y)\\ &=10(2x+2y+2z)+(2x+2y+2z)\\ &=11(2x+2y+2z)\\ &=22(x+y+z).\end{aligned}$$

We divide this sum by $(x+y+z)$, the sum of the digits of the original number, which gives us

$$\frac{22(x+y+z)}{(x+y+z)}=22.$$

This shows the value of algebra in explaining simple arithmetic phenomena.

The Special 24 Property

The number 24 has a unique property that is not well known. If you take any prime number, square it, and then subtract 1, the result will be a multiple of 24.

To exhibit this property, consider the prime number 17. We square 17 to get 289, and then subtract 1 to get 288, which is 12×24. To convince you that it could be a general rule, we will take a larger prime number such as 91. The process goes as follows:

$$91^2=8281, \text{ then } 8281-1=8280=24\times 345.$$

A motivated reader might wish to use algebra to justify this rule.

A Curious Uniqueness of the Number 26

It is hard to fathom that only one number, 26, is sandwiched in between a square number ($25 = 5^2$) and a cube number ($27 = 3^3$). To date, another number sandwiched in between a square and a cube has not been found among numbers less than 10 quadrillion. This certainly makes the number 26 unique.

The Surprise Appearance of the Number 68

The decimal system sometimes lends itself to rather unusual results. One example occurs if you select any number whose units digit is 8 and raise that number to the 5th power. The resulting number will always end in 68. Here are a few examples of this phenomenon.

$$8^5 = 327\mathbf{68}$$
$$18^5 = 1{,}889{,}5\mathbf{68}$$
$$38^5 = 79{,}235{,}1\mathbf{68}$$
$$58^5 = 656{,}356{,}7\mathbf{68}.$$
$$478^5 = 24{,}953{,}960{,}486{,}3\mathbf{68}.$$

The Attractive Cube of 71

Surprising sequences can appear coincidentally, such as when we take the cube of the number 71, which is $71^3 = 357{,}911$. Notice the figures that comprise this number: 3, 5, 7, 9, 11. Interesting!

The Magic Number 73

The process described here will conclude with 73 regardless of the number we select at the beginning. Begin with an eight-digit number with the first four digits repeated as the second four digits, such as the number 35,783,578. We divide this number by 137 to get 261,194. We then divide that number 261,194 by the original first four digits, 3578, and we will always end up with the magic number 73. Just to convince the readership that this will work for any eight-digit number (repeating four digits), we will do the procedure with the number 86,578,657: $86{,}578{,}657 \div 137 = 631{,}961$, and we then divide $631{,}961 \div 8657 = 73$, as was predicted. Try the procedure with other numbers to see this work again and again.

The Omnipresent Number 76

Take a power of any number that ends in 76; 376; 9,376; 09,376; 109,376; or 7,109,376.

The powers also end in these numbers. For example, consider these:

$$23276^4 = 293{,}517{,}090{,}687{,}774{,}9\mathbf{76}$$
$$56376^3 = 179{,}177{,}212{,}325{,}\mathbf{376}$$
$$52109376^4 = 7{,}373{,}326{,}924{,}224{,}858{,}840{,}692{,}347{,}\mathbf{109{,}376}$$
$$127{,}109{,}376^3 = 2{,}053{,}679{,}935{,}765{,}149{,}30\mathbf{7{,}109{,}376}.$$

Uniqueness of the Number 88

There are certain numbers that have a unique characteristic, such as the number 88. The number 88 is the *only* two-digit number whose square consists of two pairs of equal digits. That is, $88^2 = 7{,}744$.

A Strange Characteristic of the Number 91

Consider a number consisting of two three-digit equal numbers, such as 247,247. The amazing characteristic of such a number is that it always is a multiple of 91. That is,

$$\frac{247{,}247}{91} = 2717.$$

This can also be seen with some other randomly selected numbers, such as $843{,}843 \div 91 = 9273$, and $473{,}473 \div 91 = 5203$.

The Unique Prime in the 90s

Another uniqueness with the numbers in the base 10 system is that there is only one group of ten numbers less than 100 that includes exactly one prime number. These are the numbers 90–99, where the number 97 is the only prime. That makes the number 97 unique amongst the ten numbers in the 90s.

A Genuine Curiosity of 101

When two numbers have a sum of 101, the difference of the squares of these two numbers is equal to the difference of the two numbers repeated. Such is the case when we consider the number 64 and 37, whose sum is 101, as we can see in the following:

$$64 + 37 = 101$$
$$64^2 - 37^2 = 4096 - 1369 = 27\mathbf{27}$$
$$64 - 37 = \mathbf{27}.$$

This is true for any two numbers who sum is 101. They need not be two-digit numbers, we as show below:

$$93 + 8 = 101$$
$$93^2 - 8^2 = 8649 - 64 = 85\mathbf{85}$$
$$93 - 8 = \mathbf{85}.$$

A Coincidence with Powers of the Number 103

The strangest things can happen with numbers that are not particularly popular. This is the case with the number 103. When we take any *odd* power of 103 and add 1, the resulting number is a multiple of 13, and if we subtract 1 from any odd power of 103, the resulting number is a multiple of 17.

For example, $103^3 = 1{,}092{,}727$. When we increase that by 1, we get $1{,}092{,}728 = 13 \times 84{,}056$.

When we decrease this power of 103 by 1, we get $1{,}092{,}726 = 17 \times 64{,}278$.

Another example could be to take $103^5 = 11{,}592{,}740{,}743$, and add 1 to get $11{,}592{,}740{,}744 = 13 \times 891{,}749{,}288$. When we subtract 1 from this power of 103, we get $11{,}592{,}740{,}742 = 17 \times 681{,}925{,}926$. Once again, we find that the power of 103 is either 1 less than a multiple of 13, or 1 greater than a multiple of 17.

However, when we take 103 to an *even* power and subtract one, that number has both 13 and 17 as factors as we can see with the following example.

$103^4 = 112{,}550{,}881$, and when we subtract 1, we get $112{,}550{,}880 = 13 \times 17 \times 509{,}280$, showing that both 13 and 17 are factors of $103^4 - 1$.

The Surprising Number 120

We consider any prime number greater than 5 and raise it to the fourth power. When we subtract one from the number, the result will always be multiple of 120.

Take, for example, the prime number 31 raised to the fourth power: $31^4 = 923{,}521$. We then subtract 1 from that number and get $923{,}520 = 120 \times 7696$.

To secure your understanding, we can repeat the process with another prime number greater than 5, such as 37: $37^4 = 1{,}874{,}161$, we subtract 1 from that number and get $1{,}874{,}160 = 120 \times 15{,}618$.

Coincidences with Prime Number 131

You will recall that a prime number has only two factors: the number itself and the number 1. The number 131 is a prime number, its digits are prime numbers, and its digits can be rearranged in any order and only generate prime numbers. Furthermore, the sum of the digits, $1 + 3 + 1 = 5$, is also prime number.

The number 131 can also be expressed as a sum of three consecutive prime numbers as follows: $131 = 41 + 43 + 47$. Perhaps even more impressively, the number 131 can be expressed as follows:

$$131 = (1^0 + 3^0 + 1^0) + (1^1 + 3^1 + 1^1) + (1^2 + 3^2 + 1^2) + (1^3 + 3^3 + 1^3) + (1^4 + 3^4 + 1^4).$$

Moreover, the sum of the nonprime numbers from 4 through the prime number 19 is equal to 131, that is, $131 = 4 + 6 + 8 + 9 + 10 + 12 + 14 + 15 + 16 + 18 + 19$. A further curiosity regarding the number 131 is that it is the sum of the odd numbers that have no even digits from 1 to 31, as shown here: $131 = 1 + 3 + 5 + 7 + 9 + 11 + 13 + 15 + 17 + 19 + 31$.

If we place a 7 on either side of 131, we get the number 71,317, which is a palindrome and also a prime number. The same thing is true if we tag a 1 to each side of the number 131 to get 11,311, which again is both a palindrome and a prime number.

A truly amazing characteristic of the number 131 can be seen with the fraction $\frac{1}{131}$. The reciprocal of the number 131 also offers a surprise. $\frac{1}{131} \approx 0.00763358778625954198473282442748091603053$ $4351145038167938931297709923664122137404580152671755$7 $2519083969465648854961832061068702 29$... and these 130 digits then repeat indefinitely.

The Number 666—The Number of the Beast

The number 666 is often referred to as "the number of the beast." One possible explanation is that 666 is associated with the Roman emperor Nero (r. 54 – 68 C.E.), the first Roman emperor to persecute Christians. His name, when written out in Aramaic and subjected to the Hebrew numerology of gematria, brings you to the number 666. However, mathematics has some interesting aspects related to the number 666, as presented here:

- The sum of the squares of the first seven prime numbers, that is, $2^3 + 3^2 + 5^2 + 7^2 + 11^2 + 13^2 + 17^2 = 666$. Also, the number 666 is the sum of the first 36 natural numbers:

$$1 + 2 + 3 + \ldots + 34 + 35 + 36 = 666.$$

- Even a palindromic relationship can generate the number of the beast:

$$1^3 + 2^3 + 3^3 + 4^3 + 5^3 + 6^3 + 5^3 + 4^3 + 3^3 + 2^3 + 1^3 = 666.$$

- The sum of the digits of 666 $(6 + 6 + 6 = 18)$ is equal to the sum of the digits of its prime factors. Its prime factors are $666 = 2 \times 3 \times 3 \times 37$, and the sum of their digits is $2 + 3 + 3 + 3 + 7 = 18 = 6 + 6 + 6$.
- Two consecutive triangular numbers, 15 and 21, also play a role with the number 666, in that $15^2 + 21^2 = 225 + 441 = 666$.
- The number 666, in a rather creative way, can create a prime number as follows: $666 \times 66 \times 6 + 1 = 263{,}737$.
- Notice the role 6s display in this equation:

$$666 = 6^3 + 6^3 + 6^3 + 6 + 6 + 6.$$

Another repetition of 6s to consider is $666 = 1^6 - 2^6 + 3^6$.

- The first gap of 666 numbers between two pairs of twin prime numbers (consecutive prime numbers) occurs between the pair of twin prime numbers 774,131 and 774,133, and the next pair of twin prime numbers 774,797, and 774799. That is, 774,797 – 774,131 = 666, and 774,799 – 774,133 = 666.
- With the help of the number 666, we can get a pair of twin prime numbers as follows: 666 + 666 + 666 + 1 = 1999, and 666 + 666 + 666 – 1 = 1997.
- Here are a few delectable number relationships, which lead to 666:

$$666 = 1^6 - 2^6 + 3^6$$
$$666 = (6+6+6) + (6^3+6^3+6^3)$$
$$666 = (6^4 - 6^4 + 6^4) - (6^3+6^3+6^3) + (6+6+6)$$
$$666 = 5^3 + 6^3 + 7^3 - (6+6+6)$$
$$666 = \left(2^1 \times 3^2\right) + \left(2^3 \times 3^4\right).$$

- We can even generate 666 by representing each of its three digits in terms of 1, 2, and 3:

$$6 = 1+2+3$$
$$6 = 1\times 2\times 3$$
$$6 = \sqrt{1^3+2^3+3^3}.$$

Therefore,

$$666 = [(100)\times(1+2+3)] + [(10)\times(1\times 2\times 3)] + \left[\sqrt{1^3+2^3+3^3}\right].$$

- Consider the value of π to 144 places: $\pi \approx$ 3.14159265358979323 846264338327950288419716939937510582097494459230781640628620899862803482534211706798214808651328230664709384460955058223172535 9. Surprisingly, the sum of these 144 digits is 666.
- Incidentally, when written in Roman numeral form, the number 666 uses all the Roman symbols prior to the M (1,000) in descending order: 666 = DCLXVI.

- Consider the arrangements of the keys on calculator:

  7 8 9
  4 5 6
  1 2 3

  If we take any row, column, or diagonal, then take its reversal, the difference of the squares of these two numbers will be a multiple of 666. For example, consider the third column, 963, and its reversal, 369. The difference of their squares $963^2 - 369^2 = 791{,}208 = 666 \times 1188$. As another example, consider the diagonal 753 and its reversal, 357. The difference of their squares $753^2 - 357^2 = 439{,}560 = 666 \times 660$. Once again, we see that the difference of the squares is a multiple of 666. And this can be done with any of the rows, columns, or diagonals of this square arrangement of the keys on a calculator.
- There is also Pythagorean triple which embraces this magic number, and that is $666^2 = 216^2 + 630^2$.

As an add-on attraction we will consider the number 666,666, where the number of the beast is used twice. This number is also significant in geometry as it is the area of a right triangle whose sides have lengths 693, 1924, and 2045, as it has an area $\frac{1}{2}(693 \times 1924) = 666{,}666$ square units. There seems to be no limit to the amazements found in mathematics.

The Number 1089

We find embedded in the number 1089 a very curious oddity, one with which you may entertain your friends. We begin by selecting any three-digit number whose units digit and hundreds digits are not the same and then reversing that number. Now subtract the two numbers you have (obviously, the larger minus the smaller). Add to this difference the reversal of its digits. Quite unexpectedly the result will always be 1089.

To ensure that you understand and appreciate this wonder, we will randomly select a three-digit number, say 732. We now subtract $732 - 237 = 495$. Reversing the digits of 495 we get 594, and now we add these last two numbers: $495 + 594 = 1089$. Yes, this will hold true for all such three-digit numbers—amazing! This is a cute little "trick" that can be justified with simple algebra.

We shall represent the arbitrarily selected three-digit number, $\overline{htu}$, as $100h + 10t + u$, where h represents the hundreds digit, t represents the tens digit, and u represents the units digit. The number with the digits reversed is $100u + 10t + h$. We will let $h > u$, which would be the case either in the number you selected or its reverse. In the subtraction, $u - h < 0$; therefore, take 1 from the tens place (of the minuend) making the units place $10 + u$. Since the tens digits of the two numbers to be subtracted are equal, and 1 was taken from the tens digit of the minuend, then the value of this digit is $10(t - 1)$. The hundreds digit of the minuend is $h - 1$, because 1 was taken away to enable subtraction in the tens place, making the value of the tens digit $10(t - 1) + 100 = 10(t + 9)$. When we do the first subtraction, we get:

$$\begin{array}{l} 100(h-1)+10(t+9)+(u+10) \\ \underline{100u+10t+h} \\ 100(h-u-1)+10(9)+u-h+10 \end{array}$$

Reversing the digits of this difference

$$100(h - u - 1) + 10 \times 9 + (u - h + 10)$$

gives us

$$100(u - h + 10) + 10 \times 9 + (h - u - 1).$$

Adding these last two expressions gives us

$$100(h - u - 1) + 10 \times 9 + (u - h + 10) + 100(u - h + 10) + 10 \times 9 + (h - u - 1) = 1000 + 90 - 1 = \mathbf{1089}.$$

This algebraic justification enables us to inspect the general case of this arithmetic process, allowing us to guarantee that this process holds true for all applicable numbers.

The number 1089 has some other oddities attached to it. One characteristic of this number can be seen by taking the reciprocal and getting the following: $\frac{1}{1089} = .\overline{0009182736455463728191}$. With the exception of the first three zeros and the last 1, we have a palindromic number 918,273,645,546,372,819, since it reads the same in both directions.

There are still more oddities involving the number 1089. For example, if we multiply 1089 by 5, we also get a palindromic number, 5445, and if we multiply 1089 by 9, we get 9801, the reverse of the original number.

Let us now multiply by 9 some numbers that are modifications of 1089: 10,989, 109,989, 1,099,989, 10,999,989, and so on, and then marvel at the results.

$$10989 \times 9 = 98901$$
$$109989 \times 9 = 989901$$
$$1099989 \times 9 = 9899901$$
$$10999989 \times 9 = 98999901$$
$$109999989 \times 9 = 989999901$$

and so on. This will continue no matter how many 9s are in the original number as long as it begins with 10 and ends with 89.

Another oddity involving the number 1089 is that $1089 = 33^2 = 65^2 - 56^2$. This is the only two-digit number that displays this pattern.

Let's look at the first nine multiples of 1089:

$$10891 = 1089$$
$$10892 = 2178$$
$$10893 = 3267$$
$$10894 = 4356$$
$$10895 = 5445$$
$$10896 = 6534$$
$$10897 = 7623$$
$$10898 = 8712$$
$$10899 = 9801.$$

Do you notice a pattern among the products? Look at the first and ninth products, that is, 1089 and 9801, which are the reverses of one another. The second and the eighth products 2178 and 8712 are also reverses of one another. And so, the pattern continues until the fifth product, 5445, is the reverse of itself, a palindromic number. The only other number of four or fewer digits whose multiple is the reverse of the original number is 2178, since $2178 \times 4 = 8712$, which just happens to be 2×1089. Wouldn't it be nice if we could extend this as we did with the above example, and insert 9s into the middle of the number to generate other numbers that have the same property. Here we have it!

$$21978 \times 4 = 87912$$
$$219978 \times 4 = 879912$$
$$2199978 \times 4 = 8799912$$
$$21999978 \times 4 = 87999912$$
$$219999978 \times 4 = 879999912$$
$$2199999978 \times 4 = 8799999912$$

and so on.

As if the number 1089 didn't already have enough cute properties, here is another one. We will consider the number 1089 in two parts, the numbers 1 and 89.

Let's see what happens when you take any number, find the sum of the squares of its digits, and repeat, continuing to add the squares of the digits of each new number. Each time, curiously enough, you will eventually reach 1 or 89. Take a look at some examples that follow.

We will begin with the number 30, and we will find the sum of the digits of this number:

$3^2 + 0^2 = 9$, $9^2 = 81$, $8^2 + 1^2 = 65$, $6^2 + 5^2 = 61$, $6^2 + 1^2 = 37$,
$3^2 + 7^2 = 58$, $5^2 + 8^2 = \mathbf{89}$,
$8^2 + 9^2 = 145$, $1^2 + 4^2 + 5^2 = 42$,
$4^2 + 2^2 = 20$, $2^2 + 0^2 = 4$, $4^2 = 16$, $1^2 + 6^2 = 37$, $3^2 + 7^2 = 58$,
$5^2 + 8^2 = \mathbf{89}$, **...**

Once we reach 89, we get into a loop when we repeat the process, since we always seem to return to the number 89. Let's try this with the number 31:

$$3^2 + 1^2 = 10,\ 1^2 + 0^2 = \mathbf{1},\ 1^2 = \mathbf{1}.$$

Again, for the number 1 a loop is formed, getting us back to 1 over and over.

We shall now try 32:

$$3^2 + 2^2 = 13,\ 1^2 + 3^2 = 10,\ 1^2 + 0^2 = \mathbf{1},\ 1^2 = \mathbf{1}.$$

We shall now try 33:

$$3^2 + 3^2 = 18,\ 1^2 + 8^2 = 65,\ 6^2 + 5^2 = 61,\ 6^2 + 1^2 = 37,$$
$$3^2 + 7^2 = 58,\ 5^2 + 8^2 = \mathbf{89},\ 8^2 + 9^2 = 145,\ 1^2 + 4^2 + 5^2 = 42,$$
$$4^2 + 2^2 = 20,\ 2^2 + 0^2 = 4,\ 4^2 = 16,\ 1^2 + 6^2 = 37,\ 3^2 + 7^2 = 58,$$
$$5^2 + 8^2 = \mathbf{89}, \ldots$$

Once again, we reached the number 89 and it repeats indefinitely.

For the number 80 as a starting number:

$$8^2 + 0^2 = 64,\ 6^2 + 4^2 = 52,\ 5^2 + 2^2 = 29,\ 2^2 + 9^2 = 85,$$
$$8^2 + 5^2 = \mathbf{89},\ 8^2 + 9^2 = 145,\ 1^2 + 4^2 + 5^2 = 42,$$
$$4^2 + 2^2 = 20,\ 2^2 + 0^2 = 4,\ 4^2 = 16,$$
$$1^2 + 6^2 = 37,\ 3^2 + 7^2 = 58,$$
$$5^2 + 8^2 = \mathbf{89}, \ldots$$

This will continue for every number we select, as you can see in the following examples:

Starting with 81:

$$8^2 + 1^2 = 65,\ 6^2 + 5^2 = 61,\ 6^2 + 1^2 = 37,\ 3^2 + 7^2 = 58,$$
$$5^2 + 8^2 = \mathbf{89},\ 8^2 + 9^2 = 145,\ 1^2 + 4^2 + 5^2 = 42,\ 4^2 + 2^2 = 20,$$
$$2^2 + 0^2 = 4,\ 4^2 = 16,\ 1^2 + 6^2 = 37,\ 3^2 + 7^2 = 58,\ 5^2 + 8^2 = \mathbf{89}, \ldots$$

However, with the number 82 we also have a repeat, but this time it is with the 1.

Using 82:

$$8^2 + 2^2 = 68,\ 6^2 + 8^2 = 100,\ 1^2 + 0^2 + 0^2 = \mathbf{1},\ 1^2 = \mathbf{1}.$$

Then for 85 we get:

$$8^2 + 5^2 = \mathbf{89},\ 8^2 + 9^2 = 145,\ 1^2 + 4^2 + 5^2 = 42,\ 4^2 + 2^2 = 20,$$
$$2^2 + 0^2 = 4,\ 4^2 = 16,\ 1^2 + 6^2 = 37,\ 3^2 + 7^2 = 58,\ 5^2 + 8^2 = \mathbf{89}, \ldots$$

The Perennial Number Scheme

This scheme will produce any one of the following numbers depending upon the size of the number we begin with: 10,989, 109,989, 1,099,989, 10,999,989, etc.

We will begin with the simplest case and select any five-digit number where the first and last digits differ by more than 1. Suppose we choose the number 83,654. We then reverse the first and last digits to get the number 43,658 and then subtract the two numbers:

$$83{,}654 - 43{,}658 = 39{,}996.$$

Once again, we interchange the first and last digits to get the number 69,993, which we then add the previous number to get $39{,}996 + 69{,}993 = 109{,}989$, which is the second number on the list above. Had we begun with a four-digit number, we would have gotten as a result 10,989. Had we started with a six-digit number, the result would have been 1,099,989.

To firm up this procedure, we shall repeat it with a six-digit number: 361,857.

- Interchange the first and last digits and subtract: $761{,}853 - 361{,}857 = 399{,}996$.
- Once again, interchange the first and last digits of the number and add: $699{,}993 + 399{,}996 = 1{,}099{,}989$.

Unusual Numbers

There are numbers that, for basically inexplicable reasons, produce some amazing results. You may be surprised at the results you will produce with some of these unusual numbers. We shall begin with the number 76,923, which we will multiply sequentially by the numbers 1, 10, 9, 12, 3, and 4. Look at the results of these six multiplications, and you will see the most amazing relationship among them.

$$
\begin{aligned}
76{,}923 \times 1 &= \mathbf{0}76{,}923 \\
76{,}923 \times 10 &= \mathbf{7}69{,}230 \\
76{,}923 \times 9 &= \mathbf{6}92{,}307 \\
76{,}923 \times 12 &= \mathbf{9}23{,}076 \\
76{,}923 \times 3 &= \mathbf{2}30{,}769 \\
76{,}923 \times 4 &= \mathbf{3}07{,}692.
\end{aligned}
$$

Notice how the first digits of the numbers generated show the original number 076,923. Furthermore, notice how the numerals are sequenced in each of the products presented. If you were to draw a diagonal line through these products from the upper right to the lower left, it would consist of the number 3. If you inspect the products further, you will notice patterns formed by parallels to the diagonal just highlighted.

The number 76,923, when multiplied by 2, 7, 5, 11, 6 and 8, provides another striking pattern.

$$
\begin{aligned}
76{,}923 \times 2 &= \mathbf{1}53{,}846 \\
76{,}923 \times 7 &= \mathbf{5}38{,}461 \\
76{,}923 \times 5 &= \mathbf{3}84{,}615 \\
76{,}923 \times 11 &= \mathbf{8}46{,}153 \\
76{,}923 \times 6 &= \mathbf{4}61{,}538 \\
76{,}923 \times 8 &= \mathbf{6}15{,}384.
\end{aligned}
$$

The first digits of these products form the same number as the first product, namely, 153,846. Here as well, the diagonal from the upper right to the left of the products is consists of 6s, and parallel lines are also represented by the same digits.

Another number that yields some symmetric results is obtained when we divide 999,999 by 7 to get 142,857. This time, we will multiply the number 142,857 sequentially by 1, 3, 2, 6, 4, and 5 to get the following:

$$142{,}857 \times 1 = 142{,}857$$
$$142{,}857 \times 3 = 428{,}571$$
$$142{,}857 \times 2 = 285{,}714$$
$$142{,}857 \times 6 = 857{,}142$$
$$142{,}857 \times 4 = 571{,}428$$
$$142{,}857 \times 5 = 714{,}285.$$

Again, you will see the first digits of each of the products to be 1, 4, 2, 8, 5, and 7, and the diagonal from upper right to lower left consists of all 7s. Similar patterns, as previously shown, are embedded in this group of products. Such unusual number patterns further exhibit the wonders that can be found in mathematics.

The Peculiar Number 90,625

The number 90,625 has a peculiar characteristic. When the number is taken to various powers, its last five digits always will be 90,625, as we can see with the following:

$$90{,}625^2 = 8{,}212{,}890{,}625$$
$$90{,}625^3 = 744{,}293{,}212{,}890{,}625$$
$$90{,}625^4 = 67{,}451{,}572{,}418{,}212{,}890{,}625$$
$$90{,}625^5 = 6{,}112{,}798{,}750{,}400{,}543{,}212{,}890{,}625$$
$$90{,}625^7 = 50{,}203{,}747{,}549{,}676{,}336{,}348{,}056{,}793{,}212{,}890{,}625$$
$$90{,}625^8 = 4{,}549{,}714{,}621{,}689{,}417{,}981{,}542{,}646{,}884{,}918{,}212{,}890{,}625$$
$$90{,}625^9 = 412{,}317{,}887{,}590{,}603{,}504{,}577{,}302{,}373{,}945{,}713{,}043{,}212{,}890{,}625$$
$$90{,}625^{10} = 37{,}366{,}308{,}562{,}898{,}442{,}602{,}318{,}027{,}638{,}830{,}244{,}541{,}168{,}212{,}890{,}625$$
$$90{,}625^{11} = 3{,}386{,}321{,}713{,}512{,}671{,}360{,}835{,}071{,}254{,}768{,}990{,}911{,}543{,}369{,}293{,}212{,}890{,}625$$
$$90{,}625^{12} = 306{,}885{,}405{,}287{,}085{,}842{,}075{,}678{,}332{,}463{,}439{,}801{,}358{,}617{,}842{,}197{,}418{,}212{,}890{,}625.$$

This pattern continues on.

An Unusual Number Curiosity

In Figure 1-16, we demonstrate a rather unusual relationship. Notice the middle digit of each of the numbers on the left and the numerators on the right. Also notice the relationship between the numbers on the left and the denominators on the right.

121	=	$\frac{22 \times 22}{1+2+1}$
12321	=	$\frac{333 \times 333}{1+2+3+2+1}$
1234321	=	$\frac{4444 \times 4444}{1+2+3+4+3+2+1}$
123454321	=	$\frac{55555 \times 55555}{1+2+3+4+5+4+3+2+1}$
12345654321	=	$\frac{666666 \times 666666}{1+2+3+4+5+6+5+4+3+2+1}$
1234567654321	=	$\frac{7777777 \times 7777777}{1+2+3+4+5+6+7+6+5+4+3+2+1}$
123456787654321	=	$\frac{88888888 \times 88888888}{1+2+3+4+5+6+7+8+7+6+5+4+3+2+1}$
12345678987654321	=	$\frac{999999999 \times 999999999}{1+2+3+4+5+6+7+8+9+8+7+6+5+4+3+2+1}$

Figure 1-16

Powers Generating Number Patterns

Consider the number 131 and enjoy the following pattern:

$$131^2 = 17{,}161$$
$$1331^2 = 1{,}771{,}561$$
$$13331^2 = 177{,}715{,}561$$
$$133331^2 = 17{,}777{,}155{,}561$$
$$1333331^2 = 1{,}777{,}771{,}555{,}561.$$

Generating Perfect Numbers

Not all numbers are perfect! However, there is a category of numbers that deserves the adjective "Perfect." Such a number is one which is equal to the sum of its proper factors, such as the number $6 = 1 + 2 + 3$, or the number $28 = 1 + 2 + 4 + 7 + 14$. For centuries, it has been a challenge to find larger perfect numbers. The first nine perfect numbers are provided in Figure 1-17, with the aid of a formula $(2^{n-1})(2^n - 1)$, which generates perfect numbers for the values of n, where n is a prime number and $(2^n - 1)$ is also a prime number.

n	**Perfect Numbers $(2^{n-1})(2^n - 1)$**
2	6
3	28
5	496
7	8,128
13	33,550,336
17	8,589,869,056
19	137,438,691,328
31	2,305,843,008,139,952,128
61	265,845,599,156,983,174,465,469,261,5953,842,176

Figure 1-17

At a glance, we notice that all of the perfect numbers have a units digit of either 6 or 8. Also, after the first four consecutive prime numbers, the next prime number, 11, is omitted, and the reason is quite simple. It does not conform to the requirements stated above. Specifically, when we inspect $(2^n - 1)$ for the value of $n = 11$, we get $(2^{11} - 1) = 2047 = 23 \times 89$, and is not a prime number. Therefore, the number 11 cannot be used to generate a perfect number from the given formula. The search goes on for many more perfect numbers!

The Addition Oddity of Six Numbers

This oddity is fun, as we will consider a seemingly baffling situation and then analyze the conclusion with simple algebra. We begin by selecting any three-digit number with no two digits the same and without zero as a digit. Then make five other numbers with these same digits. Suppose you have selected the number 473. The list of all the numbers formed from these three digits is

473
437
347
374
743
734

The oddity is to find the sum of these numbers without actually adding them. How can this be done so fast? All we have to do is get the sum of the digits of the original number and then multiply it by 222 to get the sum of these six numbers. Here, we get $4 + 7 + 3 = 14$, then $14 \times 222 = 3108 = 473 + 437 + 347 + 374 + 743 + 734$. You can use this nifty exercise to challenge a friend, and arrive at the sum much faster than your friend will using traditional addition.

To justify this scheme, we need to determine why we use the number 222. Let us inspect some of the many quirks of number properties using simple algebra. Consider the number $\overline{abc} = 100a + 10b + c$, where a, b, c are numbers from the set {1, 2, 3, ..., 9}. The sum of the digits is $a + b + c$. We now represent all six numbers formed by these digits as:

$$\begin{aligned} &100a + 10b + c \\ +\, &100a + 10c + b \\ +\, &100b + 10a + c \\ +\, &100b + 10c + a \\ +\, &100c + 10a + b \\ +\, &100c + 10b + a \end{aligned}$$

When we add these numbers, we get:

$$\begin{aligned} &100(2a + 2b + 2c) + 10(2a + 2b + 2c) + 1(2a + 2b + 2c) \\ &= 200(a + b + c) + 20(a + b + c) + 2(a + b + c) \\ &= 222(a + b + c), \text{ which is 222 times the sum of the digits.} \end{aligned}$$

This explains from where the number 222 emanates.

Amazing Power Relationships

When we investigate number properties we tend to stumble on some amazing oddities, such as when the sum of a number's digits are related to the number in unusual ways. Take, for example, the number 81, whose digit sum is $8 + 1 = 9$, and $9^2 = 81$, which is what we started with. The same is true with the number 4913, whose digit sum is $4 + 9 + 1 + 3 = 17$, and $17^3 = 4913$.

In both cases, we took the sum of the digits to a power and ended up with the number we started with. To find other such numbers is no mean feat. Figure 1-18 provides lots of examples of these unusual numbers. The beauty is self-evident!

Number	=	(Sum of the Digits)^n	Number	=	(Sum of the Digits)^n
81	=	9^2	34,012,224	=	18^6
			8,303,765,625	=	45^6
512	=	8^3	24,794,911,296	=	54^6
4913	=	17^3	68,719,476,736	=	64^6
5,832	=	18^3			
17,576	=	26^3	612,220,032	=	18^7
19,683	=	27^3	10,460,353,203	=	27^7
			27,512,614,111	=	31^7
2,401	=	7^4	52,523,350,144	=	34^7
234,256	=	22^4	271,818,611,107	=	43^7
390,625	=	25^4	1,174,711,139,837	=	53^7
614,656	=	28^4	2,207,984,167,552	=	58^7
1,679,616	=	36^4	6,722,988,818,432	=	68^7
17,210,368	=	28^5	20,047,612,231,936	=	46^8
52,521,875	=	35^5	72,301,961,339,136	=	54^8
60,466,176	=	36^5	248,155,780,267,521	=	63^8
205,962,976	=	46^5			

20,864,448,472,975,628,947,226,005,981,267,194,447,042,584,001 = 207^{20}

Figure 1-18

An Unusual Number Pattern

There are times when we can appreciate numbers by their relationships, which are common but often neglected. These unexpected number arrangements shown in Figure 1-19 should give you a renewed appreciation of the unusual beauty of numbers.

1	=	1	=	1×1	=	1^2
1+2+1	=	2+2	=	2×2	=	2^2
1+2+3+2+1	=	3+3+3	=	3×3	=	3^2
1+2+3+4+3+2+1	=	4+4+4+4	=	4×4	=	4^2
1+2+3+4+5+4+3+2+1	=	5+5+5+5+5	=	5×5	=	5^2
1+2+3+4+5+6+5+4+3+2+1	=	6+6+6+6+6+6	=	6×6	=	6^2
1+2+3+4+5+6+7+6+5+4+3+2+1	=	7+7+7+7+7+7+7	=	7×7	=	7^2
1+2+3+4+5+6+7+8+7+6+5+4+3+2+1	=	8+8+8+8+8+8+8+8	=	8×8	=	8^2
1+2+3+4+5+6+7+8+9+8+7+6+5+4+3+2+1	=	9+9+9+9+9+9+9+9+9	=	9×9	=	9^2

Figure 1-19

In Figure 1-20, we find another curious relationship that ought to be appreciated.

1^3	=	1^2
$1^3 + 2^3$	=	$(1+2)^2$
$1^3 + 2^3 + 3^3$	=	$(1+2+3)^2$
$1^3 + 2^3 + 3^3 + 4^3$	=	$(1+2+3+4)^2$
$1^3 + 2^3 + 3^3 + 4^3 + 5^3$	=	$(1+2+3+4+5)^2$
$1^3 + 2^3 + 3^3 + 4^3 + 5^3 + 6^3$	=	$(1+2+3+4+5+6)^2$
$\vdots$		$\vdots$
$1^3 + 2^3 + 3^3 + 4^3 + 5^3 + 6^3 + \ldots + n^3$	=	$(1+2+3+4+5+6+\ldots+n)^2$

Figure 1-20

Now for a real treat. Consider these two sets of numbers, {1,4,6,7} and {2,3,5,8}, compare the sum of each set, and then compare the sum of their squares:

$$1 + 4 + 6 + 7 = 2 + 3 + 5 + 8 = 18$$
$$1^2 + 4^2 + 6^2 + 7^2 = 2^2 + 3^2 + 5^2 + 8^2 = 102.$$

Now we will consider the same numbers taken to the third power. At first glance, the comparison is not obvious.

$$1^3 + 4^3 + 6^3 + 7^3 = 1 + 64 + 216 + 343 = 624$$
$$2^3 + 3^3 + 5^3 + 8^3 = 8 + 27 + 125 + 512 = 672.$$

We then add 8 to each of the bases and take the third powers to get the following:

$$9^3 + 12^3 + 14^3 + 15^3 = 729 + 1728 + 2744 + 3375 = 8576$$
$$10^3 + 11^3 + 13^3 + 16^3 = 1000 + 1331 + 2197 + 4096 = 8624.$$

When we take the difference of the sums as shown below, we end up with a surprising equality: $672 - 624 = 8624 - 8576 = 48$, or we could write this as $672 + 8576 = 624 + 8624$.

This lets us conclude the following equality:

$$2^3 + 3^3 + 5^3 + 8^3 + 9^3 + 12^3 + 14^3 + 15^3$$
$$= 2^3 + 3^3 + 5^3 + 8^3 + 10^3 + 11^3 + 13^3 + 16^3.$$

There are many other relationships that can be drawn from the above which will be left to the motivated reader.

Considering Powered Numbers

The famous Swiss mathematician Leonhard Euler (1707–1783) proved that it is impossible to express the sum of two cubes as the cube of another number. We can come close, insofar as $9^3 + 10^3 = 1729$, which is close to $1728 = 12^3$, and $6^3 + 8^3 = 728$, which is close to $729 = 9^3$.

There are other close cubes as shown in Figure 1-21.

$$
\begin{aligned}
&1^3 + 2^3 = 2^3 + 1\\
&9^3 + 10^3 = 12^3 + 1\\
&135^3 + 138^3 = 172^3 - 1\\
&791^3 + 812^3 = 1010^3 - 1\\
&11161^3 + 11468^3 = 14258^3 + 1
\end{aligned}
$$

Figure 1-21

Yet, the sum of two cubes can be equal to the sum of two other cubes:

$$
\begin{aligned}
1^3 + 12^3 &= 1729 = 9^3 + 10^3\\
2^3 + 16^3 &= 4104 = 9^3 + 15^3\\
2^3 + 34^3 &= 39312 = 15^3 + 33^3\\
9^3 + 34^3 &= 40033 = 16^3 + 33^3\\
10^3 + 27^3 &= 20683 = 19^3 + 24^3.
\end{aligned}
$$

A common example where a cubed number can be expressed as the sum of three other cubed numbers is the following: $6^3 = 3^3 + 4^3 + 5^3$.

Euler further conjectured that no fourth power number could equal the sum of fewer than four other fourth power numbers. This leads us to a few examples where we can express fourth power numbers as the sum of four other fourth power numbers. We could consider fifth power numbers, where we need to find the sum of five other fifth power numbers which is equal to the initial fifth power number. We show a few examples here.

$$72^5 = 67^5 + 47^5 + 46^5 + 43^5 + 19^5$$

$$107^5 = 7^5 + 43^5 + 57^5 + 80^5 + 100^5.$$

Powers as the sum of six fifth powers:

$$12^5 = 11^5 + 9^5 + 7^5 + 6^5 + 5^5 + 4^5$$

$$30^5 = 29^5 + 19^5 + 16^5 + 11^5 + 10^5 + 5^5.$$

Surprisingly, in 1966 two mathematicians, Leon J. Lander and Thomas R. Parkin, discovered a counter example to Euler's conjecture. They found that a fifth power can be expressed as the sum of fewer than five fifth powers with the following example:

$$144^5 = 27^5 + 84^5 + 110^5 + 133^5.$$

Furthermore, in 1988 another counterexample to Euler's conjecture regarding fourth powers was discovered, namely, that a fourth power can be expressed as the sum of fewer than four fourth powers, as we can see with this example:

$$422481^4 = 414560^4 + 217519^4 + 95800^4.$$

There is no limit to the experiments and discoveries that can be made with powers of numbers. Let this motivate further exploration.

Sums of Consecutive Integers

Which numbers can be expressed as the sum of consecutive integers? You may want to experiment a bit before trying to generate a rule. You might try to express the first batch of natural numbers as the sum of consecutive integers. We provide a list of some early numbers in Figure 1-22.

2 = not possible	21 = 1 + 2 + 3 + 4 + 5 + 6
3 = 1 + 2	22 = 4 + 5 + 6 + 7
4 = not possible	23 = 11 + 12
5 = 2 + 3	24 = 7 + 8 + 9
6 = 1 + 2 + 3	25 = 12 + 13
7 = 3 + 4	26 = 5 + 6 + 7 + 8
8 = not possible	27 = 8 + 9 + 10
9 = 4 + 5	28 = 1 + 2 + 3 + 4 + 5 + 6 + 7
10 = 1 + 2 + 3 + 4	29 = 14 + 15
11 = 5 + 6	30 = 4 + 5 + 6 + 7 + 8
12 = 3 + 4 + 5	31 = 15 + 16
13 = 6 + 7	32 = not possible
14 = 2 + 3 + 4 + 5	33 = 10 + 11 + 12
15 = 4 + 5 + 6	34 = 7 + 8 + 9 + 10
16 = not possible	35 = 17 + 18
17 = 8 + 9	36 = 1 + 2 + 3 + 4 + 5 + 6 + 7 + 8
18 = 5 + 6 + 7	37 = 18 + 19
19 = 9 + 10	38 = 8 + 9 + 10 + 11
20 = 2 + 3 + 4 + 5 + 6	39 = 19 + 20
	40 = 6 + 7 + 8 + 9 + 10

Figure 1-22

These numbers represented by sums of consecutive numbers are clearly not unique. For example, 30 can be expressed in more than one way: $30 = 9+10+11$, or $30 = 6+7+8+9$.

An inspection of the chart shows that a consecutive number sum cannot represent the powers of 2.

This is an interesting fact. It is not something that one would expect. By making a list of these consecutive number sums, you ought to begin to see patterns. Clearly, the triangular numbers are equal to the sum of the first n natural numbers, that is, 1, 2, 3, 4, 5,… . A multiple of 3, say $3n$, can always be represented by the sum $(n-1)+n+(n+1)$. You may discover other patterns.

For the purists and the more ambitious readers, we now prove this conjecture. First, we will establish when a number can be expressed as a sum of at least two consecutive positive integers. Let us analyze what values can be taken by the sum (S) of two (or more) consecutive positive integers from a to b, where $b > a$. We have $S = a+(a+1)+(a+2)+\cdots+(b-1)+b = \left(\frac{a+b}{2}\right)(b-a+1)$, found by applying the formula for the sum of an arithmetic series.* Then doubling both sides we get $2S = (a+b)(b-a+1)$.

Letting $(a+b) = x$ and $(b-a+1) = y$, we can note that x and y are both integers and since their sum, $x+y = 2b+1$, is odd, either x or y is odd and the other is even. Note that $2S = xy$.

Case 1. S is a power of 2.
Let $S = 2^n$. We have $2(2^n) = xy$, or $2^{n+1} = xy$. The only way we can express 2^{n+1} as a product of an even and an odd number is if the odd number is 1. If $x = a+b = 1$, then a and b cannot be positive integers. If $y = b - a + 1 = 1$, then we have $a = b$, which also cannot occur. Therefore, S cannot be a power of 2.

Case 2. S is not a power of 2.
Let $S = m2^n$, where m is an odd number greater than 1. We have $2(m2^n) = xy$, or $m2^{n+1} = xy$. We will now find positive integers a and b

*The sum of an arithmetic series is $S = \frac{n}{2}(a+l)$, where n is the number of terms, and a is the first term and l is the last term.

such that $b > a$ and $S = a + (a+1) + \cdots + b$. The two numbers 2^{n+1} and m are not equal, since one is odd and the other is even. Therefore, one is bigger than the other. Let x be the bigger one, and y be the smaller one. This assignment gives us a solution for a and b, as $x + y = 2b + 1$, giving a positive integer value for b, and $x - y = 2a - 1$, giving a positive integer value for a. Also, $y = b - a + 1 > 1$, so $b > a$, as required. We have obtained a and b. Therefore, for any S that is not a power of 2 we can find positive integers a and b, where $b > a$, such that $S = a + (a+1) + \cdots + b$.

In summary, a number can be expressed as a sum of (at least two) positive integers, if, and only if, the number is not a power of 2. This may appear harmless, but you should be aware that we have just accomplished some complicated mathematics while admiring the beauty of the subject.

Curious Pythagorean Triples

A Pythagorean triple consists of three numbers a, b, and c where $a^2 + b^2 = c^2$. Here is a most unusual sequence of Pythagorean triples where the largest member of the triple is the smallest in the next triple:

$$(3, 4, 5) \to (5, 12, 13) \to (13, 84, 85) \to$$
$$(85, 3612, 3613) \to (3613, 6{,}526{,}884, 6{,}526{,}885).$$

Not only that, but we should also notice that the second two numbers in each triple are consecutive numbers. Furthermore, the second number in each triple has a factor which is the second number in the previous triple. That is,

$$\mathbf{12} = 4 \times 3,$$
$$\mathbf{84} = \mathbf{12} \times 7,$$
$$\mathbf{3612} = \mathbf{84} \times 43,$$
$$6{,}526{,}884 = \mathbf{3612} \times 1807.$$

There doesn't seem to be a limit to the beauty that can be found amongst the Pythagorean triples.

Powers Determine Digital Oddities

It is always entertaining and rewarding when we find unusual number relationships determined by the digits they exhibit. For example, there are two numbers when squared present nine digits exactly once, as seen here: $567^2 = 321,489$ and $854^2 = 729,316$.

Furthermore, two consecutive powers of the number 69 together exhibit all ten digits exactly once: $69^2 = 4761$ and $69^3 = 328,509$. The same is true for the number 18, where consecutive powers result in two numbers that represent all ten digits: $18^3 = 5832$ and $18^4 = 10,4976$.

There are also base relationships which represent nine digits exactly once:

$$19,635 + 48 = 27^3$$
$$19,638 + 45 = 27^3$$
$$19,648 + 35 = 27^3$$
$$19,645 + 38 = 27^3.$$

Another peculiarity is square numbers that only use two digits, such as:

$$121 = 11^2, 144 = 12^2, 225 = 15^2, 400 = 20^2, 441 = 21^2,$$
$$484 = 22^2, 676 = 26^2, 900 = 30^2, 1444 = 38^2,$$
$$7744 = 88^2, 10,000 = 100^2, 11,881 = 109^2, 29,929 = 173^2,$$
$$44,944 = 212^2, 55,225 = 235^2, 9,696,996 = 3114^2.$$

A motivated reader might seek other such square numbers represented by only two digits.

The Beauty of Palindromes

Remember that palindromic numbers are those that read the same frontwards and backwards. Dates can be a source for some symmetric inspection. For example, the year 2002 is a palindrome, as is the year 1991. Moreover, those who have lived through 1991 and 2002 will be the last generation who will have lived through two palindromic years for over the next 1000 years. There were several dates in October

2001 that appeared as palindromes when written in American style: 10/1/01, or 10/22/01, and others. Europeans had the ultimate palindromic moment at 8:02 P.M. on February 20, 2002, since they would have written it as 20.02, 20-02-2002. It is a bit thought-provoking to come up with other palindromic dates.

There is a nice technique for generating palindromic numbers. One can arrive at a palindromic number by taking a given number and adding it to its reversal, continuing this process until a palindromic number is reached. There are times when this can be accomplished in very few steps. For example, if one were to begin with the number 86, the palindrome would be accomplished in three steps: 86 + 68 = 154, 154 + 451 = 605, and 605 + 506 = 1111, a palindromic number. Starting with 97 will require six steps to reach a palindrome, while the number 98 will require 24 steps to reach the palindromic number 8,813,200,023,188, which is probably the maximum number of steps it takes to achieve the palindrome.

However, the real fame that the number 196 enjoys in mathematics is that it is the smallest number that does not produce a palindromic number through the reversal process. (At least, to date it has not produced a palindromic number.) The next few numbers that do not generate a palindromic number with this procedure are: 295, 394, 493, 592, 689, 691, 788, 790, 879, 887, ...

As we have seen, a palindromic number can either be a prime number or a composite number. For example, 151 is a prime palindrome and 171 is a composite palindrome. Yet, with the exception of 11, a palindromic prime must have an odd number of digits.

There are some lovely patterns when dealing with palindromic numbers. For example, numbers that yield palindromic cubes are palindromic themselves, as can be seen with:

$$\begin{aligned}
11^3 &= 1331\\
101^3 &= 1{,}030{,}301\\
1001^3 &= 1{,}003{,}003{,}001\\
10001^3 &= 1{,}000{,}300{,}030{,}001\\
100001^3 &= 1{,}000{,}030{,}000{,}300{,}001
\end{aligned}$$

... and this pattern continues on.

Furthermore, the first four powers of 11 are palindromic numbers:

$$11^1 = 11$$
$$11^2 = 121$$
$$11^3 = 1331$$
$$11^4 = 14641.$$

The number 193 lends itself to an interesting manipulation. Consider $193^2 = 37{,}249$. If we tag onto this square the reverse number, 94,273, we get the palindrome 3,724,994,273, which is the product of palindromic prime number 3,721,273, and palindromic composite number 1001, that is, 3,721,273 × 1001 = 3,724,994,273, a palindrome! There are lots more curiosities for the ambitious reader to find amongst palindromic numbers.

A Weird Development of a Palindromic Number

There are palindromic numbers that can be generated in a very strange fashion. Here is just one example:

$$\frac{2 \cdot 22 \cdot 222 \cdot 2222}{2 \cdot 2 \cdot 2 \cdot 2} = 1356531 = 123321 \cdot 11.$$

Another example is $123{,}321 = 111 \cdot 1111$. Another strange generation of palindromes is $(1234567 \cdot 9) + 8 = 11{,}111{,}111$. Notice how the sum of three palindromes can generate another palindromic number: $101 + 131 + 151 = 383$. We also can find the smallest palindrome that is the product of three consecutive prime numbers: $505 = 5 \cdot 7 \cdot 11 \cdot 13$.

Another unexpected palindrome appears when we take the sum of the fifth powers of the first thirteen natural numbers: $1^5 + 2^5 + 3^5 + 4^5 + 5^5 + 6^5 + 7^5 + 8^5 + 9^5 + 10^5 + 11^5 + 12^5 + 13^5 = 1002001$.

Think Before Counting

Very often a problem seems so simple that we plunge right in without first considering a strategy. An impetuous beginning often leads to a less elegant solution than achieved with a bit of forethought. Here are two simple problems that can be made even simpler by thinking before working on them.

The first one is as follows: *Find all pairs of prime numbers whose sum equals 999.*

You might begin by taking a list of prime numbers and trying various pairs to see if they add up to 999. This is obviously very tedious and time-consuming, and you would never be quite certain that you had considered all pairs of prime. Let's use some logical reasoning to solve this problem. In order to obtain an odd sum for two numbers (prime or otherwise), exactly one of the numbers must be even. Since there is only one even prime, namely, 2, there can be only one pair of primes whose sum is 999, and that pair is 2 and 997. That, now, seems so simple.

A second problem where pre-planning and orderly thinking can help is as follows:

How many palindromes are there between 1 and 1000 inclusive?

The traditional approach to this problem would be to write out all the numbers between 1 and 1000, and then count which ones are palindromes. However, this is a cumbersome and time-consuming task at best, and one could easily omit some of them. Let's see if we can look for a pattern to solve the problem in a more orderly fashion.

Range	Number of Palindromes	Total Palindromes
1–9	9	9
10–99	9	18
100–199	10	28
200–299	10	38
300–399	10	48
.	.	.
.	.	.
900–999	10	108

There is a pattern. There are exactly 10 palindromes in each group of 100 numbers (after 99). Thus, there will be nine sets of 10, or 90, plus the 18 from numbers 1 to 99, for a total of 108 palindromes between 1 and 1000.

Another solution to this problem would involve organizing the data in a favorable way. Consider all the single-digit numbers (self-palindromes). There are 9. There are also 9 two-digit palindromes. The three-digit palindromes have 9 possible "outside digits" and 10 possible "middle digits," so there are 90 three-digit palindromes. In total there are 108 palindromes between 1 and 1000, inclusive. The motto is to think first, then begin to solve.

A Special Palindrome 101

The palindromic number 101 is the 26th prime number. The reciprocal of 101 generates a rather beautiful and on expected number pattern, that is

$$\frac{1}{101} \approx 0.00990099009900990099009900990099009900990099,$$

which is quite unique among prime numbers. The number 101 can be expressed as a sum of five consecutive prime numbers: $101 = 13 + 17 + 19 + 23 + 29$. The number 101 can also be expressed in terms of factorials as $101 = 5! - 4! + 3! - 2! + 1!$. Most unusually, we find that the number 101 can be obtained by taking the sum of the products of all combinations of the first four prime numbers:

$$101 = (2 \times 3) + (5 \times 2) + (2 \times 7) + (3 \times 5) + (3 \times 7) + (5 + 7).$$

Furthermore, the number 101 is the largest known prime that is 1 greater than a power of 10.

There is a clever test to determine if a number is divisible by 101. Although it is a bit cumbersome, it does provide some insight into the number 101. Going from right to left, separate the number into pairs of digits, and then alternate between subtracting and adding the pairs. At the end, either a 0 or 101 would indicate that the original number is divisible by 101. For example, let's determine if the number 854,089,043,463 is divisible by 101. Begin by separating the number into pairs of digits from right to left: 63, 34, 04, 89, 40, 85.

85, 40, 89, 04, 34, 63. Then subtract and add pairs alternately: $63 - 34 + 04 - 89 + 40 - 85 = - 101$. Therefore, the original number was divisible by 101.

The number 101 is the first of 21 prime numbers where, when the center digit is removed, the remaining number is still a prime number. For example, here are a few of these prime numbers which convert to a prime number when the center digit is removed:

(101, 11), (103, 13), (107, 17), ... (193, 13), (197, 17) and (199, 19).

If we list the prime numbers in descending order, beginning with the number 97, we create this multi-digit number: 97 89 83 79 73 71 67 61 59 53 47 43 41 37 31 29 23 19 17 13 11 7 5 3 2. When we subtract 101 from this number, the result is a prime number:

$$9{,}789{,}837{,}973{,}716{,}761{,}595{,}347{,}434{,}137{,}312{,}923{,}191{,}713{,}117{,}532 - 101 = 9{,}789{,}837{,}973{,}716{,}761{,}595{,}347{,}434{,}137{,}312{,}923{,}191{,}713{,}117{,}431.$$

An Unusual Palindrome

At first glance, the palindromic number 134,757,431 does not show any special features. However, this number can be represented in three different ways, where the bases and exponents use the numbers 1–9 exactly once, as shown below.

$$1^7+2^3+3^8+4^5+5^4+6^2+7^1+8^9+9^6=134{,}757{,}431$$
$$1^7+2^8+3^4+4^2+5^3+6^5+7^1+8^9+9^6=134{,}757{,}431$$
$$1^7+2^5+3^8+4^1+5^2+6^4+7^3+8^9+9^6=134{,}757{,}431.$$

Notice, in each of the representations the bases are the numbers 1 through 9, and the exponents similarly use the numbers 1–9 exactly once.

When Algebra Is Not Helpful

Algebra can be a very powerful tool in mathematics. However, there are times when an algebraic solution to a problem is not an advantage. Consider the challenge: *Find four consecutive* ***numbers*** *whose product is 120.*

Give yourself a bit of time to begin to tackle this question. Most folks will probably write an algebraic equation to depict the situation. It may look like this: $x(x+1)(x+2)(x+3)=120$.

Removing the parentheses leaves us with a fourth-degree equation (a quartic equation) in one variable, which is $x^4 + 3x^3 + 2x^2 - 40 = 0$. Rather than trying to solve this quartic equation, a non-algebraic solution might be preferable. Simply guess intelligently and check to get the solution: $2 \times 3 \times 4 \times 5 = 120$. You can see from this demonstration that although algebra is very useful to introduce or explain some arithmetic relationships, it is not always the best method. This may come as a shock for many, but the beauty of mathematics is to know, through practice, when to use the techniques available to us.

Admire Friendly Numbers

Mathematicians have decided that two numbers are considered friendly (or as sometimes used in the more sophisticated literature, "amicable") if the sum of the proper divisors (which are all the divisors or factors of the number except the number itself) of one number equals the second number, *and* the sum of the proper divisors of the second number equals the first number as well. Let's look at the smallest pair of friendly numbers, 220 and 284 and see why they are friendly. The divisors (or factors) of **220** are 1, 2, 4, 5, 10, 11, 20, 22, 44, 55, and 110. Their sum is 1 + 2 + 4 + 5 + 10 + 11 + 20 + 22 + 44 + 55 + 110 = **284**. The divisors of **284** are 1, 2, 4, 71, and 142, and their sum is 1 + 2 + 4 + 71 + 142 = **220**. This shows that the two numbers can be considered *friendly numbers.* This pair of amicable numbers was already known to ancient Greek mathematician Pythagoras about 500 BCE.

A second pair of friendly or amicable numbers is 17,296 and 18,416. This pair was discovered by the French mathematician Pierre de Fermat (1601–1665), although there is evidence that this discovery was anticipated by the Moroccan mathematician Ibn al-Banna al-Marrakushi al-Azdi (1256–c.1321). To justify this amicability, we must determine the prime factors of the two numbers: $17{,}296 = 2^4 \cdot 23 \cdot 47$, and $18{,}416 = 2^4 \cdot 1151$.

The sum of the factors of 17,296 is

$$1+2+4+8+16+23+46+47+92+94+184+188+368+376+752+ \\ 1081+2162+4324+8648=\underline{18{,}416}.$$

The sum of the factors of 18,416 is

$$1+2+4+8+16+1151+2302+4604+9208=\underline{17,296}.$$

Therefore, 17,296 and 18,416 are amicable, or friendly, numbers.

The French mathematician, René Descartes (1596–1650) found another pair of amicable numbers: 9,363,584, and 9,437,056. By 1747, the Swiss mathematician Leonhard Euler (1707–1783) had discovered 60 pairs of amicable numbers, yet he seems to have overlooked the second smallest pair, namely, 1184 and 1210, which was discovered in 1866 by the 16-year-old B. Nicolò I. Paganini (1850–?). The sum of the factors of 1184 is 1 + 2 + 4 + 8 + 16 + 32 + 37 + 74 + 148 + 296 + 592 = 1210. And the sum of the factors of 1210 is 1 + 2 + 5 + 10 + 11 + 22 + 55 + 110 + 121 + 242 + 605 = 1184.

To date, over 363,000 pairs of amicable numbers have identified, although we do not know if there are an infinite number of such pairs.

Here are a few more pairs of friendly numbers:

2620 and 2924
5020 and 5564
6232 and 6368
10,744 and 10,856
12,285 and 14,595
63,020 and 76,084
66,928 and 66,992
111,448,537,712 and 118,853,793,424.

You might want to verify the above pairs' "friendliness!"

And for the expert, the following is one method for finding pairs of friendly numbers:

$$\text{Let } a=3\cdot 2^n-1$$
$$b=3\cdot 2^{n-1}-1$$
$$c=3^2\cdot 2^{2n-1}-1,$$

where n is an integer ≥ 2, and a, b and c are all prime numbers, then $2^n ab$ and $2^n c$ are friendly numbers.

A Divisibility Conundrum

We seek the smallest number that, when divided by each of the integers 2, 3, 4, 5, 6, 7, 8, 9, and 10, will yield the remainder which is 1 less than the divisor.

Bypassing a trial-and-error procedure, we will find the least common multiple of the numbers 2, 3, 4, 5, 6, 7, 8, 9, and 10. This is done by considering the factors of these numbers and noting that the least common multiple is $2^3 \cdot 3^2 \cdot 5 \cdot 7 = 2520$. We know that when this number 2520 is divided by 2, 3, 4, 5, 6, 7, 8, 9, and 10, there is no remainder. Let's consider $2520 - 1 = 2519$. When 2519 is divided by 2, 3, 4, 5, 6, 7, 8, 9, and 10, the remainder will be one less than the divisor.

Numerical Challenges

The task here is to write the number 1 by using each of the 10 digits just once.

One solution is the following:

$$1 = \frac{35}{70} + \frac{148}{296}.$$

Are there any other such solutions?

Now we write the number 100 by using each of the 10 digits just once.

One solution is the following:

$$100 = 50 + 49\frac{1}{2} + \frac{38}{76}.$$

Are there any other such solutions?

Using the number 3 five times, represent the number 31.

This can be done as follows:

$$31 = 3^3 + \frac{3}{3} + 3.$$

Another challenge would be to represent the number 7 using exactly five 2s.

This can be done as follows:

$$7 = 2 + 2 + 2 + \frac{2}{2}.$$

Using the nine digits exactly once, create a fraction that is equal to $\frac{1}{2}$.

One possible solution is the following:

$$\frac{9{,}327}{18{,}654} = \frac{1}{2}.$$

Using the numbers 1 through 9 exactly once, create the number 99,999:

One possibility is the following: $1234 + 98{,}765 = 99{,}999$.

Show how to make the number 1000 using eight 8s.

One possibility is the following: $1000 = 888 + 88 + 8 + 8 + 8$.

Use seven 4s to represent 100.

This can be done as follows: $44 + 44 + 4 + 4 + 4 = 100$.

Now use sixteen 4s to represent 1000

One possibility is to do it as follows: $444 + 444 + 44 + 44 + 4 + 4 + 4 + 4 + 4 + 4$.

The next challenge is to write the number 24 using three equal digits, none of them being 8.

One simple solution is $22 + 2 = 24$. Can you find another solution?

Identify a number which is twice the product of its digits.

The sought-after number is 36, since $3 \times 6 = 18$, and $18 \times 2 = 36$.

Which number is triple the sum of its digits?

The sought-after number is 27, since $2 + 7 = 9$, and $9 \times 3 = 27$.

A motivated reader might choose other such challenges.

Discovering a Square Number

This question will give you better insight into numbers. What is the smallest number that when multiplied by 1859 would produce a square number?

In order to come up with the smallest number that produces a perfect square when multiplied by 1859, one needs to look at the factors of $1859 = 2^4 \times 3^2 \times 11$. The only factor that is not a perfect square is the number 11. Therefore, if we multiply $1{,}859 \times 11 = 20{,}449$, which is a perfect square, namely, $\sqrt{20{,}449} = 143$.

Guess the Magic Number

We seek a number less than 500 that is a perfect square and when flipped upside down is also a perfect square.

To find this number we have to first consider which numbers can be flipped over and still be a proper number. That is, we will consider the numbers 0, 1, 6, 8, and 9. The numbers 0, 1, and 8 can be flipped over and remain with the same value. However, when we flip 9 it becomes 6, and vice versa. The possible perfect squares less than 500 using these digits are 1, 9, 16, 81, 100, 169, or 196. The only perfect square that remains a perfect square when flipped over is 169, since 961 is also perfect square, namely, $31 \times 31 = 961$. Notice the peculiar relationship $13 \times 13 = 169$.

Square Numbers

One could produce four square numbers using each of the nine digits exactly once.

The four perfect squares are 9, 81, 324, and 576, which equal $3^2, 9^2, 18^2$, and 24^2, respectively. The challenge now is to find the smallest perfect square using all nine digits exactly once, and the largest perfect square using all nine digits exactly once. The smallest perfect square under these conditions is $11{,}826^2 = 139{,}854{,}276$. The largest perfect square using all nine digits exactly once is $30{,}384^2 = 923{,}187{,}456$.

While we are dealing with largest possible numbers, let's refer to the rule for checking divisibility by 11, where we established that the difference of the odd placed digits and the even placed digits has to be a multiple of 11 for the number to be a multiple of 11. This is the case with 102,347,586, since the odd placed digits have a sum of $1 + 2 + 4 + 5 + 6 = 18$, and the even placed digits have a sum of $0 + 3 + 7 + 8 = 18$; the difference is 0, which is clearly a multiple of 11. And now we come to the largest number that satisfies divisibility by 11, namely, 987,652,413. Here the odd placed digits have a sum of $9 + 7 + 5 + 4 + 3 = 28$, the even placed digits have a sum of $8 + 6 + 2 + 1 = 17$, and the difference of these two sums is $28 - 17 = 11$, which is clearly a multiple of 1. Therefore, the original number is a multiple of 11. And it is claimed that 987,652,413 is the largest number that has this property, as

$$\frac{987,652,413}{11} = 89,786,583.$$

A Multiplication Puzzle Using Numbers 1 to 9

The challenge here is to create multiplication examples where all the digits from 1 to 9 are used. Nine possibilities are presented below.

$4 \times 1738 = 6952$ $\quad 4 \times 1963 = 7852$ $\quad 12 \times 483 = 5796$
$18 \times 297 = 5346$ $\quad 27 \times 198 = 5346$ $\quad 28 \times 157 = 4396$
$39 \times 186 = 7254$ $\quad 42 \times 138 = 5796$ $\quad 48 \times 159 = 9632$

Curious Arithmetic

Multiplication presents some curiosities that are not easily explained. There are times when we multiply a larger number by a single-digit number, and the resulting product's units digit is the leftmost digit of the larger multiplicand. One such example is when we multiply **1**42857 × 3 = 42857**1**. Another such example uses a 16-digit number that we multiply by 3 and then divide by 2, resulting in a number where the leftmost digit of the original largest factor is slotted into the units digit: **3**,529,411,746,705,882 × 3 = 10,588,235,240,117,646 ÷ 2 = 5,294,117,620,058,82**3**. Although these may give you some insight into arithmetic, most would consider these functions simple entertainment, which we need to discover in mathematics.

Strange Equalities

When we speak of unusual or strange relationships in math, the following must rank high on the list. The reader will probably want to know why this happens. What about these numbers allows this relationship to occur? There are no clever answers to these questions. All we can do is find and limit the possibilities. There are times when the numbers speak more effectively than any explanation. Here is one such case. Just look at these equalities and enjoy them for what they are: Unusual!

$$1^1+6^1+8^1=15=2^1+4^1+9^1$$
$$1^2+6^2+8^2=101=2^2+4^2+9^2$$

$$1^1+5^1+8^1+12^1=26=2^1+3^1+10^1+11^1$$
$$1^2+5^2+8^2+12^2=234=2^2+3^2+10^2+11^2$$
$$1^3+5^3+8^3+12^3=2366=2^3+3^3+10^3+11^3$$

$$1^1+5^1+8^1+12^1+18^1+19^1=63=2^1+3^1+9^1+13^1+16^1+20^1$$
$$1^2+5^2+8^2+12^2+18^2+19^2=919=2^2+3^2+9^2+13^2+16^2+20^2$$
$$1^3+5^3+8^3+12^3+18^3+19^3=15{,}057=2^3+3^3+9^3+13^3+16^3+20^3$$
$$1^4+5^4+8^4+12^4+18^4+19^4=260{,}755=2^4+3^4+9^4+13^4+16^4+20^4.$$

Not much can be said here. Perhaps "wow" would be in order. Beauty is in the unusual!

Every so often we stumble on some relatively useless information that still contains an element of beauty. One such example is the unusual relationship (encountered earlier in this chapter) between the numbers shown in Figure 1-23. Just enjoy the peculiarity and see if you can find others. What we have are pairs of numbers where the product and the sum are reversals of each other.

Pair		Product	Sum
9	9	81	18
3	24	72	27
2	47	94	49
2	497	994	499

Figure 1-23

Can you find another pair of numbers that exhibit this unusual property?

Triangular Numbers

Recall that *triangular numbers* have gained their title because they are quantities (perhaps of coins or dots) that can be arranged into an equilateral triangle, such as those shown in Figure 1-24. The number of dots in each arrangement is then called a triangular number. The first several triangular numbers are 1, 3, 6, 10, 15, 21, 28, and 36.

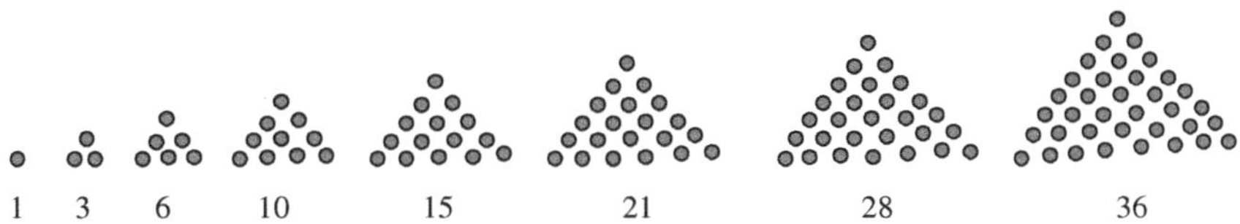

Figure 1-24

Some triangular numbers have additional features that make them extra special. For example, triangular number 6 squared is 36, which is also a triangular number. The next largest triangular number whose square is also a triangular number is a huge number with 660 digits, which is more than we can exhibit here.

We should mention that there are many triangular numbers that are also square numbers. Here are just a few:

$$0 = 0$$
$$1 = 1$$
$$36 = 6^2$$
$$1225 = 35^2$$
$$41{,}616 = 204^2$$
$$1{,}413{,}721 = 1189^2.$$

There is a curious way to generate a square number from a triangular number. We can take any triangular number, multiply it by 72, add 9, and we have a square number. For example, suppose we take the triangular number 28. Following this procedure, we get $28 \times 72 = 2016$. When we add 9 to 2016, we get $2025 = 45^2$. Another example using the triangular number 136 follows: $136 \times 72 = 9792$; we then add 9 to get $9801 = 99^2$. Once again, we end up with a square number generated from a triangular number. More in this regard follows in the next few units. (See Appendix A for a list of triangular numbers.)

A Triangular Number Curiosity

As we now investigate several characteristics and relationships of triangular numbers, it would be useful to see the first 100 triangular numbers:

> 0, 1, 3, 6, 10, 15, 21, 28, 36, 45, 55, 66, 78 66 78, 91, 105, 120, 136, 153, 171, 190, 210, 231, 253, 276, 300, 325, 351, 378, 406, 435, 465, 496, 528, 561, 595, 630, 666, 703, 741, 780, 820, 861, 903, 946, 990, 1035, 1081, 1128, 1176, 1225, 1275, 1326, 1378, 1431, 1485, 1540, 1596, 1653, 1711, 1770, 1830, 1891, 1953, 2016, 2080, 2145, 2211, 2278, 2346, 2415, 2485, 2556, 2628, 2701, 2775, 2850, 2926, 3003, 3081, 3160, 3240, 3321, 3403, 3486, 3570, 3655, 3741, 3828, 3916, 4005, 4095, 4186, 4278, 4371, 4465, 4560, 4656, 4753, 4851, 4950, 5050 . . . .

Another curiosity regarding triangular numbers is when you multiply one of these numbers by 8 and add 1 you always end up with a square number, which is a bit simpler than the previous method mentioned above. For example, suppose we take the triangular number $136 \times 8 = 1088$ then $+ 1 = 1089 = 33^2$. As another example, consider the triangular number 2016: $2016 \times 8 = 16{,}128$, then $+ 1 = 16{,}129 = 127^2$. Convince yourself by trying this relationship with other triangular numbers.

Triangular numbers have a variety of other secrets embedded in them. If you take the product of two consecutive triangular numbers, multiply that product by 16, and add 1, the result will always be square number. We can see this with the following example. Consider the two consecutive triangular numbers 66 and 78. The process dictates the following: $66 \times 78 = 5148$, we then multiply this by 16 to get 82,368 and add 1 to get $82{,}369 = 287^2$.

While we are lauding triangular numbers, we should note that perfect numbers, which we recall are numbers equal to the sum of their factors, such as 6, 28, 496, and 8128, are also triangular numbers.

Generating Squares and Triangular Numbers with Oblong Numbers

Oblong numbers are generated by the product of two consecutive numbers. For example, 12 and 30 are considered oblong numbers because they consist of two consecutive factors, that is, $3 \times 4 = 12$, and $5 \times 6 = 30$. The sequence of oblong numbers is 2, 6, 12, 20, 30, 42, 56, 72, and so on. Here are some fascinating properties of oblong, square, and triangular numbers:

- Twice a triangular number yields an oblong number. For example, $2 \times 28 = 56$, which is an oblong number, since $7 \times 8 = 56$.
- The sum of two consecutive triangular numbers yields a square number. For example, $15 + 21 = 36$, which is a square number.
- Two consecutive square numbers plus the square of an oblong number between them yields a square number. For example, $36 + 42^2 + 49 = 1849 = 43^2$.
- Two consecutive oblong numbers plus twice the square number between them yields another square number. For example, $12 + 2(16) + 20 = 64 = 8^2$.
- The sum of an oblong number and the next square number yields a triangular number. For example, $12 + 16 = 28$.
- The sum of a square number and the next oblong number yields a triangular number. For example, $16 + 20 = 36$.
- A number plus the square of that number is equal to an oblong number. For example, $6 + 36 = 42$.

But there is still more to follow about triangular numbers.

More Unusual Characteristics of Triangular Numbers

Following are some truly mind-boggling relationships between triangular numbers, square numbers, prime numbers, etc.

- Nine times a triangular number plus 1 is a triangular number. For example, $(9 \times 21) + 1 = 190$, which is a triangular number.
- Eight times a triangular number +1 is a perfect square. For example, $(8 \times 21) + 1 = 169 = 13^2$.
- The sum of n consecutive cubes beginning with one is equal to the square of the nth triangular number. That is, $T_n^2 = 1^3 + 2^3 + 3^3 + 4^3 + \cdots + n^3$. As an example, consider the fifth triangular number, 15. Therefore, we have $T_5^2 = 1^3 + 2^3 + 3^3 + 4^3 + 5^3 = 225 = 15^2$.
- A triangular number greater than 1 can never be a cube, a fourth power, or a fifth power.
- All perfect numbers are also triangular numbers, such as 6 and 28.
- Only one triangular number is a prime number, and that is number 3.
- Some triangular numbers are palindromic—that is, the same in both directions—such as the following palindromic triangular numbers: 1, 3, 6, 55, 66, 171, 595, 666, 3003, 5995, 8778, 15,051, 66,066, 617,716, 828,828, 1,269,621, 1,680,861, 3,544,453, 5,073,705, 5,676,765, 6,295,926, 351,335,153.
- Some triangular numbers have reversals that are also triangular numbers: 1, 3, 6, 10, 55, 66, 120, 153, 171, 190, 300, 351, 595, 630, 666, 820, 3003, 5995, 8778, 15,051, 17,578, 66,066, 87,571, 156,520, 180,300, 185,745, 547,581, 557,040, 617,716, 678,030, 828,828.
- There are triangular numbers that are also square numbers, such as: 1, 36, 1225, 41,616, 1,413,721, 48,024,900, 1,631,432,881, 55,420,693,056.
- There are pairs of triangular numbers whose sum and difference are also triangular numbers. Following are some examples of such pairs: (15, 21), (105, 171), (378, 703), (780, 990), (1485, 1486), (2145, 3741), (5460, 6786), (7875, 8778). The pair of triangular numbers (105, 171) yields the sum $105 + 171 = 276$, which is the 23rd triangular number, and the difference $171 - 105 = 66$ which is the 11th triangular number. Just to take another example, the pair of triangular numbers (703, 378) have the sum $703 + 378 = 1081$, which is the 46th triangular number, and the difference $703 - 378 = 325$, which is the 25th triangular number.

- There are some triangular numbers which are the product of three consecutive numbers. For example, the 15th triangular number, 120, is the product of three consecutive numbers: $5 \times 6 \times 7 = 120$. Also, consider $9 \times 10 \times 11 = 990$, which is the 44th triangular number.
- An unusual feature with the triangular number 120 is that it is the product of sets of three, four, and five consecutive numbers: $4 \times 5 \times 6 = 2 \times 3 \times 4 \times 5 = 1 \times 2 \times 3 \times 4 \times 5 = 120$.
- Naturally, there are also some triangular numbers that are the product of two consecutive numbers, such as $2 \times 3 = 6$, which is the third triangular number, or $14 \times 15 = 210$, which is the 20th triangular number. This goes on for quite a way, such as $2870 \times 2871 = 8{,}239{,}770$, which is the 4059th triangular number.
- There are even triangular numbers that are the product of two prime numbers. For example, the triangular number $6 = 2 \times 3$, both of which are prime numbers. Some other examples are $3 \times 7 = 21$, the sixth triangular number, and $7 \times 13 = 91$, the 13th triangular number. Some other triangular numbers that are the product of two prime numbers are: 10, 15, 21, 55, 91, 253, 703, 1081, 1711, 1891, 2701, 3403, 5671, 12,403, 13,861, 15,931, etc.
- Furthermore, there are triangular numbers that are divisible by the sum of their digits. The triangular number 1128 is divisible by $1 + 1 + 2 + 8 = 12$, which is a factor of 1128, since $12 \times 94 = 1128$. Other examples of such triangular numbers are: 1, 3, 6, 10, 21, 36, 45, 120, 153, 171, 190, 210, 300, 351, 378, 465, 630, 666, 780, 820, 990, 1035, etc.
- Recall the unit on happy numbers, because there are triangular numbers which are also happy numbers. For example, the triangular number 496 is a happy number because continually taking the sum of the squares of the digits of 496 reduces to 1, as the following shows:

$$4^2 + 9^2 + 6^2 = 133, 1^2 + 3^2 + 3^2 = 19, 1^2 + 9^2 = 82, 8^2 + 2^2 = 68,$$
$$6^2 + 8^2 = 100, \text{ and } 1^2 + 0^2 + 0^2 = 1.$$

For convenience, we list the first few happy triangular numbers: 1, 10, 28, 91, 190, 496, 820, 946, 1128, 1275, 2080, 2211, 2585, 3321, 4278, 8128, 8256, 8778, 9591, 9730, 11,476, etc.

- To date, only three triangular numbers have been identified as repdigit numbers: 55, 66, and 666.
- The sum of reciprocals of triangular numbers seems to converge to 2, demonstrated as follows:

$$1+\frac{1}{3}+\frac{1}{6}+\frac{1}{10}+\frac{1}{15}+\frac{1}{21}+\frac{1}{28}+\frac{1}{36}+\frac{1}{45}+\cdots=2.$$

- There are certain pairs of triangular numbers whose product is a perfect square. Here are a few examples. The product of the second and 24th triangular numbers: $3 \times 300 = 900 = 30^2$. Also, the product of the third and 48th triangular numbers $6 \times 1176 = 7056 = 84^2$. The product of the second and the 242nd triangular numbers is $3 \times 29{,}403 = 88{,}209 = 297^2$.
- Every positive integer can be written as a sum of at most three triangular numbers.
- Every fourth power greater than 1 can be written as the sum of two triangular numbers. For example, $2^4 = 16 = 6 + 10 = T_3 + T_4$, and $2^4 = 16 = 1 + 15 = T_1 + T_5$, where T_n is the nth triangular number. Four examples follow:

$$3^4 = 81 = T_8 + T_9 = T_5 + T_{11}$$
$$4^4 = 256 = T_{15} + T_{16} = T_{11} + T_{19}$$
$$5^4 = 625 = T_{24} + T_{25} = T_{19} + T_{29}$$
$$6^4 = 1296 = T_{35} + T_{36} = T_{29} + T_{41}.$$

  This continues on.

- For any natural number n, the nth triangular number T_n can be represented by the following: $T_n = 1 + 9 + 9^2 + 9^3 + 9^4 + \cdots + 9^n$. That is, the sequence of powers of 9 generates some triangle numbers, such as the first, fourth, and thirteenth triangular numbers, as can be seen here:

$$T_1 = 1$$
$$T_4 = 1 + 9 = 10$$
$$T_{13} = 1 + 9 + 9^2 = 91$$
$$T_{40} = 1 + 9 + 9^2 + 9^3 = 820$$
$$T_{121} = 1 + 9 + 9^2 + 9^3 + 9^4 = 7381.$$

- There are some unexpected relationships among the triangular numbers as shown below:

$$T_1 + T_2 + T_3 = T_4$$
$$T_5 + T_6 + T_7 + T_8 = T_9 + T_{10}$$
$$T_{11} + T_{12} + T_{13} + T_{14} + T_{15} = T_{16} + T_{17} + T_{18}$$
$$T_{19} + T_{20} + T_{21} + T_{22} + T_{23} + T_{24} = T_{25} + T_{26} + T_{27} + T_{28}.$$

- Triangular numbers also lend themselves to some identities, such as:

$$T_n^2 = T_n + (T_{n-1} \times T_{n+1})$$
$$T_{n^2-1} = 2 \times T_n \times T_{n-1}.$$

- There is even a Pythagorean triple comprised of three triangular numbers:

$$(T_{132})^2 + (T_{143})^2 = (T_{164})^2$$
$$8778^2 + 10{,}296^2 = 13{,}530^2.$$

- Some cute triangular number arrangements are as follows:

$$T_{10} = T_{5+5} = 55$$
$$T_{100} = T_{50+50} = 5050$$
$$T_{1000} = T_{500+500} = 500500$$
$$T_{10000} = T_{5000+5000} = 50005000.$$

The pattern continues indefinitely.

- For the final amazing property of triangular numbers, notice how the position of two triangular numbers is related to their sum, as can be seen from the following examples:

$$T_{90} + T_{415} = 90{,}415$$
$$T_{585} + T_{910} = 585{,}910$$
$$T_{120} + T_{1545} = 120{,}1545$$
$$T_{150} + T_{1726} = 150{,}1726$$
$$T_{244} + T_{2196} = 2{,}442{,}196$$

Notice the relationship between the subscripts of the triangular numbers and the total value.

An Unexpected Result

There are times when a result in mathematics seems illogical and refutable. This is especially true when you have a sequence of common numbers and you can easily detect the next number, such as the list beginning with the even numbers **1, 2, 4, 8, 16**. Common logic would indicate that the next number should be 32. However, when the next number is given as **31** (instead of the expected 32), cries of "wrong!" are usually heard. Much to your amazement, this can be shown to be a legitimate sequence: **1, 2, 4, 8, 16, 31**.

The task now is to justify the legitimacy of this sequence. It would be nice if it could be justified geometrically, as that would give convincing physical evidence. We will do that later, but in the meantime, let us first find the succeeding numbers in this "curious" sequence.

We shall set up a table of differences. Figure 1-25 is a chart showing the differences between terms of the given sequence up to 31. Then working backwards, we progress until a pattern is established, as we can see at the third difference, where we have 1, 2, 3.

Original Sequence	**1**		**2**		**4**		**8**		**16**		**31**
First Difference		1		2		4		8		15	
Second Difference			1		2		4		7		
Third Difference				1		2		3			
Fourth difference					**1**		**1**				

Figure 1-25

With the fourth differences forming a sequence of constants, we can reverse the process—turn the table upside down—and extend the third differences a few more steps with 4 and 5, as shown in Figure 1-26.

We obtained the bold numbers by working backwards from the third difference sequence. Thus, the next numbers of the given

Fourth Difference					1		1		1		1				
Third Difference				1		2		3		**4**		**5**			
Second Difference			1		2		4		7		**11**		**16**		
First Difference		1		2		4		8		**15**		**26**		**42**	
Original Sequence	1		2		4		8		16		**31**		**57**		**99**

Figure 1-26

sequence are 57 and 99. The general term is a fourth power expression since we had to go to the third differences to get a constant. The general term (n) is:

$$\frac{n^4 - 6n^3 + 23n^2 - 18n + 24}{24}.$$

One should not think that this sequence is independent of other parts of mathematics. For example, consider the Pascal triangle, named after the French mathematician Blaise Pascal (1623–1662), which is shown in Figure 1-27. The Pascal triangle is a triangular arrangement of numbers beginning on top with 1, then the second row has 1, 1, then the third row is obtained by placing 1s at the end and adding the two numbers in the second row 1 + 1 = 2 to get the middle number. The fourth row is obtained the same way. After the end 1s are placed, the 3s are the sum of the two numbers above and to the right and left, that is, 1 + 2 = 3, and 2 + 1 = 3.

1
1 1
1 2 1
1 3 3 1
1 4 6 4 1
1 5 10 10 5 1
1 6 15 20 15 6 1
1 7 21 35 35 21 7 1
1 8 28 56 70 56 28 8 1

Figure 1-27

Consider the horizontal sums of the rows of the Pascal triangle to the right of the bold line: 1, 2, 4, 8, 16, 31, 57, 99, 163. This is our newly developed sequence.

A geometric interpretation should help convince you of the beauty and consistency inherent in mathematics. To do this, Figure 1-28 is a chart of the number of regions into which a circle can be partitioned by joining points on the circle. Figure 1-28 shows a circle with six points and lines joining the six points, partitioning the circle into 31 regions, as indicated by the dots in the figure.

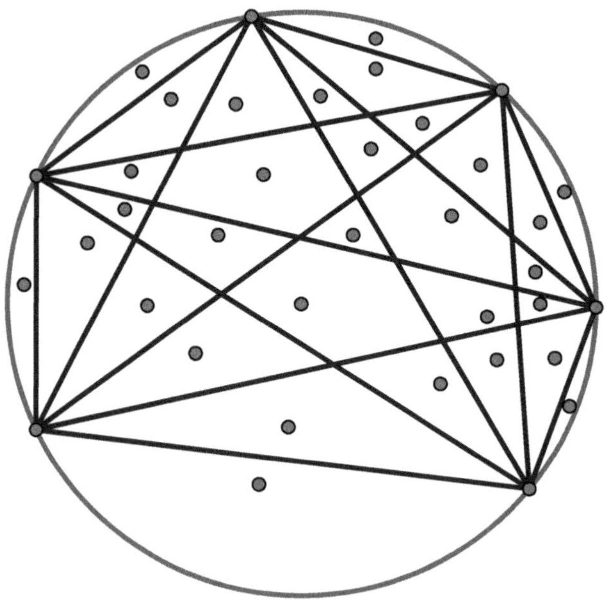

Figure 1-28

Figure 1-29 summarizes how the given number of lines partitions the circle into a certain number of regions.

Number of Points on the Circle	Number of Regions into Which the Circle Is Partitioned
1	1
2	2
3	4
4	8
5	16
6	31
7	57
8	99

Figure 1-29

Now that you see that this unusual sequence appears in various other fields, you may feel a degree of satisfaction emerging, as we have justified a strange sequence.

The Famous Ratio π

One of the most famous ratios in mathematics is the ratio of a circle's circumference to its diameter, known as π. Early estimates of the ratio of π were very crude. The ancients approximated $\pi = 3$, far from the current computer-assisted value of approximately $\pi \approx 3.1415926535\ 8979323846\ 2643383279\ 5028841971\ 6939937510\ 5820974944...$

Some very clever attempts to calculate value π came up with the following, which is rather close to the actual value:

$$\pi \approx \left(9^2 + \frac{19^2}{22}\right)^{\frac{1}{4}} = 3.14159265262...$$

Yet, perhaps the most accurate approximation of π before the invention of the computer was that of the Chinese astronomer and mathematician Zu Chongzhi (429–500), who through various mysterious ways came up with $\pi = \frac{355}{113} = 3.14159292035398230088495575\allowbreak 22123893805309734513274336283185840707964601769911504\allowbreak 424778761061946902654867256637168$

1415929203539823008849557522123893805309
7345132743362831858407079646017699115044
24778761061946902654867256637168

which continues on for 112 places before it repeats again.

What the Bible Has as the Value of π

Let's stay with this "crude" approximation of $\pi = 3$ for a moment. For centuries scholars believed that 3 was the value of π in biblical times. Virtually all the books on the history of mathematics stated that in its earliest manifestation in history, namely, in the Old Testament of the Bible, the value of π is given as 3. Yet recent "detective work" shows otherwise.[3]

One always relishes the notion that a hidden code can reveal long lost secrets. Such is the case with the common interpretation of the value of π in the Bible. There are two places in the Bible where the same sentence appears, identical in every way except for one word, which is spelled differently in the two citations. The description of a pool, or fountain, in King Solomon's temple may be found in 1 *Kings* 7:23 and 2 *Chronicles* 4:2, and reads as follows:

> *"And he made the molten sea[4] of ten cubits from brim to brim, round in compass, and the height thereof was five cubits; and a line of thirty cubits did compass it round about."*

[3] Alfred S. Posamentier and Noam Gordon. "An Astounding Revelation on the History of π." *Mathematics Teacher*, Vol. 77, No. 1, Jan. 1984, p. 52.

[4] The "molten sea" was a gigantic bronze vessel for ritual ablutions in the court of the First Temple (966–955 B.C.). It was supported on the backs of 12 bronze oxen and its volume ≈ 45,000 liters.

The circular structure described here had a circumference of 30 cubits[5] and a diameter of 10 cubits. From this, scholars concluded that the Bible has $\pi = \frac{30}{10} = 3$. This is obviously a very primitive approximation of π. Elijah of Vilna[6] (1720–1797) was a rabbi and one of the great modern biblical scholars who earned the title "*Gaon* (brilliance) of Vilna." He made a remarkable discovery that refuted claims that the Bible approximated the value of π as 3. Elijah of Vilna noticed that the Hebrew word for "line measure" was written differently in each of the two Biblical passages mentioned above.

In 1 *Kings* 7:23, "line measure" is written as קוה, whereas in 2 *Chronicles* 4:2 it is written as קו. Elijah applied the ancient biblical analysis technique (still used by Talmudic scholars today) called *gematria*, where the Hebrew letters are given numerical values according to their sequence in the Hebrew alphabet, to the two spellings of the word for "line measure" and found the following. The letter values are: ק = 100, ו = 6, and ה = 5. Therefore, the gematria value for the spelling of "line measure" in 1 *Kings* 7:23 is קוה $= 5 + 6 + 100 = 111$, while the value of the spelling in 2 *Chronicles* 4:2 is קו $= 6 + 100 = 106$. Elijah then took the ratio of these two values: $\frac{111}{106} = 1.0472$ (to four decimal places), which he considered the necessary "correction factor." By multiplying the Bible's apparent value (3) of π by this "correction factor," one gets 3.1416, which is π correct to four decimal places! "Wow!" is a common reaction. Such accuracy is quite astonishing for ancient times. Moreover, remember how just getting $\pi = 3.14$ using string measurements was quite a feat. Now imagine getting p accurate to four decimal places. We would contend that this would be nearly impossible with typical string measurements. Try it if you need convincing.

---

[5] A cubit is the length of a person's fingertip to his elbow.

[6] In those days Vilna was in Poland, while today the town is named Vilnius and is in Lithuania.

The Famous Pascal Triangle

The Pascal triangle has lots of treasures beyond what was shown in the previous section. In Figure 1-30, we see just a few of them. Horizontally, the digits determine the powers of 11, while the sums of the numbers give you the powers of 2. We also notice the sequence of natural numbers and the sequence of triangular numbers. The Fibonacci numbers are also apparent in the triangle. There is lots more for the reader to discover in the vast arrangement of numbers in the Pascal triangle.

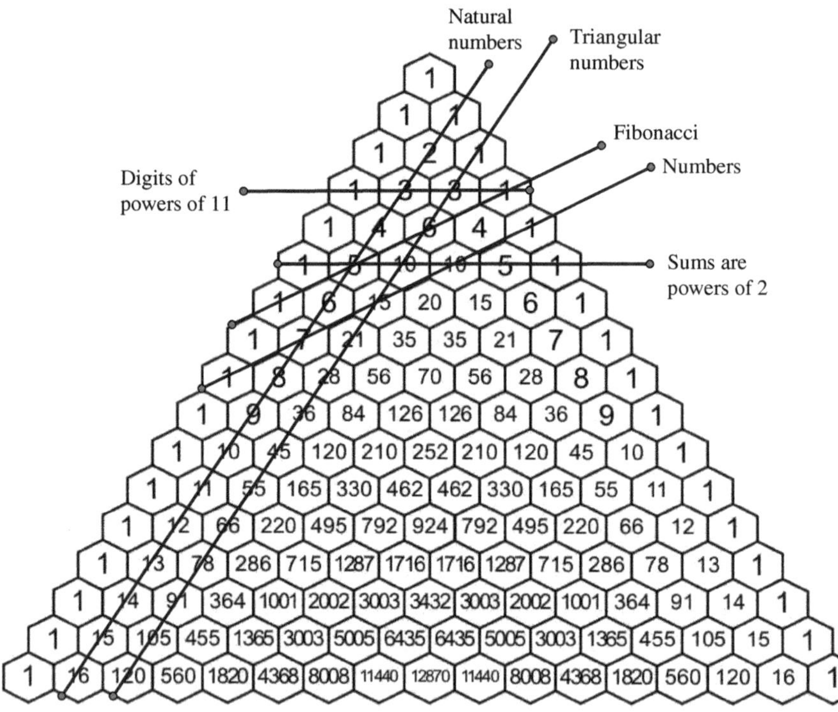

Figure 1-30

The Unusual Armstrong Numbers

As we continue to explore some of the most celebrated numbers in mathematics, we come to those that are referred to as *Armstrong numbers*, named after the American mathematician Michael F. Armstrong (1941–2020). Sometimes referred to as "narcissistic numbers," the Armstrong numbers are equal to the sum of their digits when each digit is taken to the power equal to the number of digits in the original number. For example, we have encountered the three-digit Armstrong number $153 = 1^3 + 5^3 + 3^3 = 1 + 125 + 27$.

A nine-digit Armstrong number is

$$472{,}335{,}975 = 4^9 + 7^9 + 2^9 + 3^9 + 3^9 + 5^9 + 9^9 + 7^9 + 5^9.$$

The Armstrong numbers are shown in Figure 1-31, and there are only 89 Armstrong numbers in the decimal system found to date. The largest Armstrong number is 39 digits long, and it is equal to the sum of its digits, each of which is taken to the 39th power:

$$1^{39} + 1^{39} + 5^{39} + 1^{39} + 3^{39} + 2^{39} + 2^{39} + 1^{39} + 9^{39} + 0^{39} + 1^{39} + 8^{39} + 7^{39} + 6^{39} + 3^{39} + 9^{39} + 9^{39} + 2^{39} + 5^{39} + 6^{39} + 5^{39} + 0^{39} + 9^{39} + 5^{39} + 5^{39} + 9^{39} + 7^{39} + 9^{39} + 7^{39} + 3^{39} + 9^{39} + 7^{39} + 1^{39} + 5^{39} + 2^{39} + 2^{39} + 4^{39} + 0^{39} + 1^{39} = 115{,}132{,}219{,}018{,}763{,}992{,}565{,}095{,}597{,}973{,}971{,}522{,}401.$$

The following is a list of the *consecutive* Armstrong numbers, where k is the number of digits.

$k = 3$:	370; 371;
$k = 8$:	24,678,050; 24,678,051;
$k = 11$:	32,164,049,650; 32,164,049,651;
$k = 16$:	4,338,281,769,391,370; 4,338,281,769,391,371;
$k = 25$:	3,706,907,995,955,475,988,644,380; 3,706,907,995, 955,475,988,644,381;
$k = 29$:	19,008,174,136,254,279,995,012,734,740; 19,008, 174,136,254,279,995,012,734,741;
$k = 33$:	186,709,961,001,538,790,100,634,132,976,990; 186, 709,961,001,538,790,100,634,132,976,991;
$k = 39$:	115,132,219,018,763,992,565,095,597,973,971,522, 400; 115,132,219,018,763,992,565,095,597,973,971, 522,401.

Incidentally, our first Armstrong number, 153, has some further amazing properties. It is a triangular number, and the sum of the first 17 numbers equals 153: $1 + 2 + 3 + 4 + 5 + 6 + 7 + 8 + 9 + 10 + 11 + 12 + 13 + 14 + 15 + 16 + 17 = 153$.

No.	Digits	Armstrong Number	No.	Digits	Armstrong Number
0	1	0	45	17	35641594208964132
1	1	1	46	17	35875699062250035
2	1	2	47	19	1517841543307505039
3	1	3	48	19	3289582984443187032
4	1	4	49	19	4498128791164624869
5	1	5	50	19	4929273885928088826
6	1	6	51	20	63105425988599693916
7	1	7	52	21	128468643043731391252
8	1	8	53	21	449177399146038697307
9	1	9	54	23	21887696841122916288858
10	3	153	55	23	27879694893054074471405
11	3	370	56	23	27907865009977052567814
12	3	371	57	23	28361281321319229463398
13	3	407	58	23	35452590104031691935943
14	4	1634	59	24	174088005938065293023722
15	4	8208	60	24	188451485447897896036875
16	4	9474	61	24	239313664430041569350093
17	5	54748	62	25	1550475334214501539088894
18	5	92727	63	25	1553242162893771850669378
19	5	93084	64	25	3706907995955475988644380
20	6	548834	65	25	3706907995955475988644381
21	7	1741725	66	25	4422095118095899619457938
22	7	4210818	67	27	121204998563613372405438066
23	7	9800817	68	27	121270696006801314328439376
24	7	9926315	69	27	128851796696487777842012787
25	8	24678050	70	27	174650464499531377631639254
26	8	24678051	71	27	177265453171792792366489765
27	8	88593477	72	29	14607640612971980372614873089
28	9	146511208	73	29	19008174136254279995012734740
29	9	472335975	74	29	19008174136254279995012734741
30	9	534494836	75	29	23866716435523975980390369295
31	9	912985153	76	31	1145037275765491025924292050346
32	10	4679307774	77	31	1927890457142960697580636236639
33	11	32164049650	78	31	2309092682616190307509695338915
34	11	32164049651	79	32	17333509997782249308725103962772
35	11	40028394225	80	33	186709961001538790100634132976990
36	11	42678290603	81	33	186709961001538790100634132976991
37	11	44708635679	82	34	1122763285329372541592822900204593
38	11	49388550606	83	35	12639369517103790328947807201478392
39	11	82693916578	84	35	12679937780272278566303885594196922
40	11	94204591914	85	37	1219167219625434121569735803609966019
41	14	28116440335967	86	38	12815792078366059955099770545296129367
42	16	4338281769391370	87	39	115132219018763992565095597973971522400
43	16	4338281769391371	88	39	115132219018763992565095597973971522401
44	17	21897142587612075			

Table of Armstrong Numbers

Figure 1-31

The number 153 is not only equal to the sum of the cubes of its digits, but it can also be expressed as the sum of consecutive factorials: $1! + 2! + 3! + 4! + 5! = 153$. This shows we can always find new exciting relationships with numbers.

Happy and Unhappy Numbers

It is not clear how the term *happy number* evolved. A common story attributes the name to British mathematician Reg Allenby's daughter, who came home from school after recognizing this unusual relationship between numbers and so called them happy numbers. We say that a number is considered a *happy number* if the sum of the squares of the digits defines a second number, and then the sum of the squares of the digits of that second number defines a third number, and then the sum of the squares of the digits of the third number defines a fourth number, and this continues until the sum of the squares of the digits equals the number 1. If 1 can be reached through this process, then we refer to this original number as a happy number. The numbers that do *not* eventually end with the number 1 when this scheme is applied are called *unhappy numbers*. These latter numbers will eventually end up in a loop, where they will continue through a cycle.

Let's consider one of the happy numbers, say 13, and follow this process of taking the sum of the squares of the digits continuously.

$$1^2 + 3^2 = 10$$
$$1^2 + 0^2 = 1.$$

Let's consider a slightly longer path to the number 1, by using 19.

$$1^2 + 9^2 = 82$$
$$8^2 + 2^2 = 68$$
$$6^2 + 8^2 = 100$$
$$1^2 + 0^2 + 0^2 = 1$$

Following is a list of the happy numbers from 1 to 1000. You might want to try a few of these to test for their "happiness."

1, 7, 10, 13, 19, 23, 28, 31, 32, 44, 49, 68, 70, 79, 82, 86, 91, 94, 97, 100, 103, 109, 129, 130, 133, 139, 167, 176, 188, 190, 192, 193, 203, 208, 219, 226, 230, 236, 239, 262, 263, 280, 291, 293, 301, 302, 310, 313, 319, 320, 326, 329, 331, 338, 356, 362, 365, 367, 368, 376, 379, 383, 386, 391, 392, 397, 404, 409, 440, 446, 464, 469, 478, 487, 490, 496, 536, 556, 563, 565, 566, 608, 617, 622, 623, 632, 635, 637, 638, 644, 649, 653, 655, 656, 665, 671, 673,

680, 683, 694, 700, 709, 716, 736, 739, 748, 761, 763, 784, 790, 793, 802, 806, 818, 820, 833, 836, 847, 860, 863, 874, 881, 888, 899, 901, 904, 907, 910, 912, 913, 921, 923, 931, 932, 937, 940, 946, 964, 970, 973, 989, 998, 1000.

The first few pairs of consecutive happy numbers (n; $n + 1$) are: (31, 32), (129, 130), (192, 193), (262, 263), (301, 302), (319, 320), (367, 368), (391, 392), ...

When you try to discover happy numbers, you will see that many fall into similar paths as you find the sums of the squares of the digits, such as the numbers 19 and 91. Following are the unique combinations, that is, none of these numbers will take the same path as they get to the number 1. Naturally, the above list includes variations of these numbers: 1, 7, 13, 19, 23, 28, 44, 49, 68, 79, 129, 133, 139, 167, 188, 226, 236, 239, 338, 356, 367, 368, 379, 446, 469, 478, 556, 566, 888, 899.

An example of an unhappy number is 25. Notice how this procedure will lead us into a loop and never reach the number 1.

$$
\begin{aligned}
2^2+5^2&=29\\
2^2+9^2&=85\\
8^2+5^2&=89\\
8^2+9^2&=145\\
1^2+4^2+5^2&=42\\
4^2+2^2&=20\\
2^2+0^2&=4\\
4^2&=16\\
1^2+6^2&=37\\
3^2+7^2&=58\\
5^2+8^2&=89.
\end{aligned}
$$

This then begins another loop—repeating everything again with 89, which was reached in the third stage of this process.

Among the happy numbers are those that are prime numbers as well, known as *happy prime numbers*. Those numbers less than 500 are the following: 7, 13, 19, 23, 31, 79, 97, 103, 109, 139, 167, 193, 239, 263, 293, 313, 331, 367, 379, 383, 397, 409, 487.

Mathematicians have pointed out some special happy numbers. For example, the smallest happy number that exhibits all the digits 0 to 9 is 10,234,456,789, while the smallest happy number that has no zeros and yet has all the other digits is 1,234,456,789. Then there is the smallest happy number that has no zeros and is also palindromic: 13,456,789,298,765,431. When we accept the inclusion of zeros, then the smallest palindromic happy number with all the digits is 1,034,567,892,987,654,301. Thus far, the largest happy number where no digit is repeated is 986,543,210.

We recall the second and third repunit primes, r_{19} = 1,111,111, 111,111,111,111 and r_{23} = 11,111,111,111,111,111,111,111, which also happen to be happy primes.

An ambitious reader may find other special characteristics among the happy numbers.

Curious Product-Reversals

It is hard to imagine that there are a certain number-pairs that yield the same product even when both numbers are reversed. For example, $12 \times 42 = 504$, and if we reverse the digits of each of the two factors, we get $21 \times 24 = 504$. The same is true for the number pair 36 and 84, since $36 \times 84 = 3024 = 63 \times 48$.

At this point, you may wonder if this will happen with any pair of numbers. The answer is that it will only work with the following 14 pairs of numbers:

$12 \times 42 = 21 \times 24 = 504$	$12 \times 63 = 21 \times 36 = 756$
$12 \times 84 = 21 \times 48 = 1008$	$13 \times 62 = 31 \times 26 = 806$
$13 \times 93 = 31 \times 39 = 1209$	$14 \times 82 = 41 \times 28 = 1148$
$23 \times 64 = 32 \times 46 = 1472$	$23 \times 96 = 32 \times 69 = 2208$
$24 \times 63 = 42 \times 36 = 1512$	$24 \times 84 = 42 \times 48 = 2016$
$26 \times 93 = 62 \times 39 = 2418$	$34 \times 86 = 43 \times 68 = 2924$
$36 \times 84 = 63 \times 48 = 3024$	$46 \times 96 = 64 \times 69 = 4416$

A careful inspection of these 14 pairs of numbers will reveal that in each case the product of the tens digits of each pair of numbers is equal to the product of the units digits. We can justify this algebraically as follows. For the numbers z_1, z_2, z_3, and z_4, we have:

$$z_1 z_2 = (10a + b) \times (10c + d) = 100ac + 10ad + 10bc + bd,$$

and

$$z_3 z_4 = (10b + a) \times (10d + c) = 100bd + 10bc + 10ad + ac.$$

Here a, b, c, d represent any of the ten digits: 0, 1, 2, ..., 9, where $a \neq 0$ and $c \neq 0$. We would like to show that $z_1\, z_2 = z_3\, z_4$. Therefore, $100ac + 10ad + 10bc + bd = 100bd + 10bc + 10ad + ac$, then $100ac + bd = 100bd + ac$, and $99ac = 99bd$, or $ac = bd$, which is what we observed earlier.

Some Peculiar Numbers

We will now embark on a most unusual numerical phenomenon. We will begin by considering unit fractions whose denominators are a multiple of 9, and not a multiple of 2 or of 5. Some examples of these would be the following fractions:

$$\frac{1}{27}, \frac{1}{63}, \frac{1}{81}, \frac{1}{99}, \frac{1}{117}, \frac{1}{153}, \text{and } \frac{1}{171}.$$

When we convert each of these fractions to decimals form—by dividing the numerator by the denominator—we get the following decimals:

$$\frac{1}{9 \times 3} = \frac{1}{27} = 0.037037037037037037037037037037\ldots = 0.\overline{037}$$

$$\frac{1}{9 \times 7} = \frac{1}{63} = 0.015873015873015873015873\ldots = 0.\overline{015873}$$

$$\frac{1}{9 \times 9} = \frac{1}{81} = 0.012345679012345679012345679\ldots = 0.\overline{012345679}$$

$$\frac{1}{9\times 11}=\frac{1}{99}=0.010101010101010101010101010101\ldots$$
$$=0.\overline{01}$$

$$\frac{1}{9\times 13}=\frac{1}{117}=0.008547008547008547008547008547\ldots$$
$$=0.\overline{008547}$$

$$\frac{1}{9\times 17}=\frac{1}{153}=0.00653594771241830065359477124183\ldots$$
$$=0.\overline{0065359477124183}$$

$$\frac{1}{9\times 19}=\frac{1}{171}=0.005847953216374269005847953216374269\ldots$$
$$=0.\overline{005847953216374269}.$$

Thus far, converting the unit fractions into decimals has revealed nothing spectacular . What follows is truly an amazing curiosity of our number system. Follow along as we multiply the repeating digits (without the initial zeros) by the factors of 9 from these denominators and observe the marvelous pattern that emerges. For example, in the first case, we will multiply 37 by multiples of 3, since the denominator 27 was arrived at by multiplying 9 times 3. Therefore, we will multiply 37, by 3, by 6, by 9, and so on, until we reach 27.

$$\begin{aligned}
37\times 3 &= 111\\
37\times 6 &= 222\\
37\times 9 &= 333\\
37\times 12 &= 444\\
37\times 15 &= 555\\
37\times 18 &= 666\\
37\times 21 &= 777\\
37\times 24 &= 888\\
37\times 27 &= 999.
\end{aligned}$$

The next fraction's denominators were attained by taking 9 × 7. Now multiplying multiples of 7 by the repeating digits 15,873, we once again obtain an easily recognized pattern.

$$
\begin{aligned}
15{,}873 \times 7 &= 111{,}111 \\
15{,}873 \times 14 &= 222{,}222 \\
15{,}873 \times 21 &= 333{,}333 \\
15{,}873 \times 28 &= 444{,}444 \\
15{,}873 \times 35 &= 555{,}555 \\
15{,}873 \times 42 &= 666{,}666 \\
15{,}873 \times 49 &= 777{,}777 \\
15{,}873 \times 56 &= 888{,}888 \\
15{,}873 \times 63 &= 999{,}999.
\end{aligned}
$$

Continuing along, we find these amazing—yet pretty—results for the succeeding fractions listed earlier. (See Figure 1-32.)

$$
\begin{aligned}
8{,}547 \times 13 &= 111\,111 \\
8{,}547 \times 26 &= 222\,222 \\
8{,}547 \times 39 &= 333\,333 \\
8{,}547 \times 52 &= 444\,444 \\
8{,}547 \times 65 &= 555\,555 \\
8{,}547 \times 78 &= 666\,666 \\
8{,}547 \times 91 &= 777\,777 \\
8{,}547 \times 104 &= 888\,888 \\
8{,}547 \times 117 &= 999\,999
\end{aligned}
$$

$$
\begin{aligned}
65\,359\,477\,124\,183 \times 17 &= 1\,111\,111\,111\,111\,111 \\
65\,359\,477\,124\,183 \times 34 &= 2\,222\,222\,222\,222\,222 \\
65\,359\,477\,124\,183 \times 51 &= 3\,333\,333\,333\,333\,333 \\
65\,359\,477\,124\,183 \times 68 &= 4\,444\,444\,444\,444\,444 \\
65\,359\,477\,124\,183 \times 85 &= 5\,555\,555\,555\,555\,555 \\
65\,359\,477\,124\,183 \times 102 &= 6\,666\,666\,666\,666\,666 \\
65\,359\,477\,124\,183 \times 119 &= 7\,777\,777\,777\,777\,777 \\
65\,359\,477\,124\,183 \times 136 &= 8\,888\,888\,888\,888\,888 \\
65\,359\,477\,124\,183 \times 153 &= 9\,999\,999\,999\,999\,999
\end{aligned}
$$

$$
\begin{aligned}
5\,847\,953\,216\,374\,269 \times 19 &= 111\,111\,111\,111\,111\,111 \\
5\,847\,953\,216\,374\,269 \times 38 &= 222\,222\,222\,222\,222\,222 \\
5\,847\,953\,216\,374\,269 \times 57 &= 333\,333\,333\,333\,333\,333 \\
5\,847\,953\,216\,374\,269 \times 76 &= 444\,444\,444\,444\,444\,444 \\
5\,847\,953\,216\,374\,269 \times 95 &= 555\,555\,555\,555\,555\,555 \\
5\,847\,953\,216\,374\,269 \times 114 &= 666\,666\,666\,666\,666\,666 \\
5\,847\,953\,216\,374\,269 \times 133 &= 777\,777\,777\,777\,777\,777 \\
5\,847\,953\,216\,374\,269 \times 152 &= 888\,888\,888\,888\,888\,888 \\
5\,847\,953\,216\,374\,269 \times 171 &= 999\,999\,999\,999\,999\,999
\end{aligned}
$$

Figure 1-32

You may want to extend this list of fractions and verify that the pattern continues. Just to show further peculiarities, notice what happens when we reverse the order of one of our previous multiplications (the one evolving from $\frac{1}{81}$). Again, we get an unexpected and rather curious pattern!

$$
\begin{aligned}
987{,}654{,}321 \times 9 &= \mathbf{8}{,}888{,}888{,}88\mathbf{9}\\
987{,}654{,}321 \times 18 &= \mathbf{17}{,}777{,}777{,}77\mathbf{8}\\
987{,}654{,}321 \times 27 &= \mathbf{26}{,}666{,}666{,}66\mathbf{7}\\
987{,}654{,}321 \times 36 &= \mathbf{35}{,}555{,}555{,}55\mathbf{6}\\
987{,}654{,}321 \times 45 &= \mathbf{44}{,}444{,}444{,}44\mathbf{5}\\
987{,}654{,}321 \times 54 &= \mathbf{53}{,}333{,}333{,}33\mathbf{4}\\
987{,}654{,}321 \times 63 &= \mathbf{62}{,}222{,}222{,}22\mathbf{3}\\
987{,}654{,}321 \times 72 &= \mathbf{71}{,}111{,}111{,}11\mathbf{2}\\
987{,}654{,}321 \times 81 &= \mathbf{80}{,}000{,}000{,}00\mathbf{1}.
\end{aligned}
$$

There are times when we can arrange numbers—legitimately—to create a rather unexpected pattern, as is the case when we find the decimal equivalent of the fraction

$$\frac{1}{729} = 0.\overline{001371742112482853223593964334705075445816186556927297668038408779149519890260631}.$$

The 81-digit decimal equivalent can be arranged in nine groups of nine as follows:

001 371 742
112 482 853
223 593 964
334 705 075
445 816 186
556 927 297
668 038 408
779 149 519
890 260 631.

Looking up and down the columns of this arrangement will reveal quite a few unusual patterns, such as consecutive numbers. The curiosities in arithmetic seem boundless!

Special Split Denominators

Recall that prime numbers are numbers that have no factors other than themselves and 1. Now we will consider the fractions whose denominators are prime numbers (excluding 2 and 5), and where the decimal expansion yields an even number of repeating digits. These will furnish further material to discover amazing patterns. Some of these are as follows:

$$\frac{1}{7} = .142857142857142857\ldots = .\overline{142857}$$

$$\frac{1}{11} = .090909\ldots = .\overline{09}$$

$$\frac{1}{13} = .\overline{076923}$$

$$\frac{1}{17} = .\overline{0588235294117647}$$

$$\frac{1}{19} = .\overline{052631578947368421}$$

$$\frac{1}{23} = 0.\overline{0434782608695652173913}.$$

We will now treat each one of these repeating periods as a number, split the even-digit sequence of digits into two parts, and then add them. An amazing result appears, as shown in Figure 1-33.

$\frac{1}{7} = 0.\overline{142857}$	142 857 999
$\frac{1}{11} = 0.\overline{09}$	0 9 9
$\frac{1}{13} = 0.\overline{076923}$	076 923 999
$\frac{1}{17} = 0.\overline{0588235294117647}$	05882352 94117647 99999999
$\frac{1}{19} = 0.\overline{052631578947368421}$	052631578 947368421 999999999
$\frac{1}{23} = 0.\overline{0434782608695652173913}$	04347826086 95652173913 99999999999

Figure 1-33

By the way, speaking of prime numbers, the numbers 139 and 149 are the first two prime numbers that differ by 10.

Kaprekar Numbers

Kaprekar numbers are named after their discoverer, Indian mathematician Dattaraya Ramchandra Kaprekar (1905–1986). A Kaprekar number is a number that when you take the square of the number, partition it into two parts, and the sum of the two parts equals the original number. For example, the number 297 is a Kaprekar number, since $297^2 = 88{,}209$, and when split into two parts, $88 + 209 = 297$. Some other Kaprekar numbers are shown in the table in Figure 1-34.

Kaprekar Numbers	Square of the Numbers			Decomposition of the Numbers
1	1^2	=	1	1 = 1
9	9^2	=	81	8 + 1 = 9
45	45^2	=	2,025	20 + 25 = 45
55	55^2	=	3,025	30 + 25 = 55
99	99^2	=	9,801	98 + 01 = 99
297	297^2	=	88,209	88 + 209 = 297
703	703^2	=	494,209	494 + 209 =703
999	999^2	=	998,001	998 + 001 = 999
2,223	$2{,}223^2$	=	4,941,729	494 + 1,729 = 2,223
2,728	2728^2	=	7,441,984	744 + 1,984 = 2,728
4,879	$4{,}879^2$	=	23,804,641	238 + 04,641 = 4,879
4,950	$4{,}950^2$	=	24,502,500	2,450 + 2,500 = 4,950
5,050	$5{,}050^2$	=	25,502,500	2,550 + 2,500 = 5,050
5,292	$5{,}292^2$	=	28,005,264	28 + 005,264 = 5,292
7,272	$7{,}272^2$	=	52,881,984	5,288 + 1,984 = 7,272
7,777	$7{,}777^2$	=	60,481,729	6,048 + 1,729 = 7,777
9,999	999^2	=	99,980,001	9,998 + 0,001 = 9,999
17,344	$17{,}344^2$	=	300,814,336	3,008 + 14,336 = 17,344
22,222	$22{,}222^2$	=	493,817,284	4,938 + 17,284 = 22,222
38,962	$38{,}962^2$	=	1,518,037,444	1518+37444 = 38,962
77,778	$77{,}778^2$	=	6,049,417,284	60,494 + 17,284 = 77,778

Figure 1-34

Other Kaprekar numbers are: 82,656; 95,121; 99,999; 142,857; ...; 538,461; 857,143; ....

We also have Kaprekar triples, which behave as follows: we cube the number, split the result into three parts, and the sum of those three numbers equals the original number. A simple case would be the number 8, where $8^3 = 512$, which when partitioned into three parts yields $5 + 1 + 2 = 8$. Another example would be 45: $45^3 = 91{,}125 = 9 + 11 + 25 = 45$. Some other Kaprekar triples are 10, 297, and 2322. Kaprekar number 297 is also a Kaprekar triple because $297^3 = 26{,}198{,}073$, and $26 + 198 + 073 = 297$.

It should also be noted that the smallest ten-digit Kaprekar number is 1,111,111,111, which is a repunit number we encountered earlier. Notice that $1{,}111{,}111{,}111^2 = 1{,}234{,}567{,}900{,}987{,}654{,}321$, where the sum of its parts is $123{,}456{,}790 + 0{,}987{,}654{,}321 = 1{,}111{,}111{,}111$. Furthermore, the number 22,222,222,222,222 is also a Kaprekar number since $22{,}222{,}222{,}222{,}222^2 = 493{,}827{,}160{,}493{,}817{,}283{,}950{,}617{,}284$. Then $4{,}938{,}271{,}604{,}938 + 17{,}283{,}950{,}617{,}284 = 22{,}222{,}222{,}222{,}222$.

A motivated reader can search for other Kaprekar numbers.

The Strange Powers of Certain Numbers

Consider the number 376 taken to a series of powers, as can be seen in Figure 1-35. Notice how the last three digits of each power are the same, 376.

		376^k
376^1	=	**376**
376^2	=	141,**376**
376^3	=	53,157,**376**
376^4	=	19,987,173,**376**
376^5	=	7,515,177,189,**376**
376^6	=	2,825,706,623,205,**376**
376^7	=	1,062,465,690,325,221,**376**
376^8	=	399,487,099,562,283,237,**376**
376^9	=	150,207,149,435,418,497,253,**376**
376^{10}	=	56,477,888,187,717,354,967,269,**376**
736^{11}	=	21,235,685,958,581,725,467,693,285,**376**
⋮		

Figure 1-35

The same arrangement is also found in the powers of 625, as can be seen in Figure 1-36. Again, the powers end in the same three digits, 625.

625^k

625^1	=	**625**
625^2	=	390,**625**
625^3	=	244,140,**625**
625^4	=	152,587,890,**625**
625^5	=	95,367,431,640,**625**
625^6	=	59,604,644,775,390,**625**
625^7	=	37,252,902,984,619,140,**625**
625^8	=	23,283,064,365,386,962,890,**625**
625^9	=	14,551,915,228,366,851,806,640,**625**
625^{10}	=	9,094,947,017,729,282,379,150,390,**625**
625^{11}	=	5,684,341,886,080,801,486,968,994,140,**625**
⋮		

Figure 1-36

Tricky Questions

Given the series of numbers shown in Figure 1-37, what could be the missing number?

16	06	68	88	??	98

Figure 1-37

As tricky as this question seems to be, it can be very easily solved by simply rotating the page and looking at the numbers upside-down, which would be: 91, 90, 89, 88, **87,** 86. Unfortunately, we tend to be drilled to look at a situation in a specific way, and are discouraged from being flexible.

In the same spirit as the above challenge, consider the order of numbers in the following sequence: 8, 5, 4, 9, 1, 7, 6, 3, 2, 0. If you try to look for a mathematical explanation for the number order, you will probably be thoroughly frustrated. Just as the previous problem bypassed mathematics, so does this one. If you look at the sequence and spell out the words for the numbers, you will find that they're in alphabetical order: **E**ight, **F**ive, **F**our, **N**ine, **O**ne, **S**even, **S**ix, **T**hree, **T**wo, **Z**ero.

It may be silly, but it makes the reader just a little bit more aware of problem solving in real life.

Large Curious Numbers

The challenge here is to find the largest n-digit number that is also an nth power.

It turns out that the number must be less than 10 because 10^n will have $n + 1$ digits. Therefore, the answer we seek is $9^{21} = 109{,}418{,}989{,}131{,}512{,}359{,}209$, which has 21 digits.

Pythagorean Quadruples

Most people are aware of popular Pythagorean triples where $a^2 + b^2 = c^2$, such as (3, 4, 5), (5, 12, 13), (7, 24, 25), (8, 15, 17), (12, 35, 37), and lots more. (See Appendix B.) However, it is more challenging to find Pythagorean quadruples, where $a^2 + b^2 + c^2 = d^2$.

This is can be an easy trick to solve. For example, when we have $3^2 + 4^2 = 5^2$ and $5^2 + 12^2 = 13^2$, we can substitute the first equation for 5^2 in the second equation to get $3^2 + 4^2 + 12^2 = 13^2$. The same can be done with the following two Pythagorean triples: $5^2 + 12^2 = 13^2$ and $13^2 + 84^2 = 85^2$. By substituting 13^2 in the second equation with first equation, we get $5^2 + 12^2 + 84^2 = 85^2$. There are lots more pairs of Pythagorean triples that combine to create Pythagorean quadruples, such as 8, 15, 17 and 17, 144, 145.

Prime Pairs

There are very few prime numbers in our number system that continue to be prime even when inverted. Such is the case with the prime numbers 19 and 61, which are inverts of one another. There were only two other prime number pairs that fit this peculiarity: 109 and 601, and 1609 and 6091.

Prime pairs can also be defined as those whose digits are consecutive prime numbers, such as the prime number 23, which consists of consecutive numbers 2 and 3. There are two more two-digit numbers that satisfy this definition of prime pairs: 67 and 89. There are no three-digit prime numbers consisting of three consecutive digits. There is one four-digit prime number consisting of consecutive digits and that is the number 4567. The next largest prime consisting of consecutive digits is the prime number 23,456,789.

Number Reversals

One can have fun searching for relationships between numbers and their reversals. Consider $12^2 = 144$. When we reverse this number to get 441, we find that it is another perfect square, namely, $\sqrt{441} = 21^2$, and 21 is the reversal of 12, the number we began with. A motivated reader might search for similar relationships, such as with the number 13, where $13^2 = 169$, and the reversal $31^2 = 961$, also a reversal from the previous square 169.

A Property of Twin Primes

Recall that two consecutive prime numbers are referred to as twin primes. Twin primes after 3 and 5 have the special property that their product plus 1 is a square number. Consider as an example the twin primes 17 and 19. Their product increased by 1 is $(17 \times 19) + 1 = 324$, which is a perfect square since $18^2 = 324$. To secure this relationship, consider the two twin primes 57 and 59. Following this procedure we get $(57 \times 59) + 1 = 3364 = 58^2$, which is the square number we sought.

There is another curious property involving twin primes. When you take the sum of two twin prime numbers (other than the number 3) and square the result, you will have produced a number that is a multiple of 36. For example, consider the twin primes 41 and 43. Following procedure described above gives us $(41 + 49)^2 = 90^2 = 8100 = 36 \times 225$, and so we have a multiple of 36.

The number of relationships generated by twin primes seems to be boundless. If we select a pair of twin primes (not including 3), add 58 to the smaller prime, square the sum, and subtract 9, we will get a multiple of 36, which can be partitioned into a product of consecutive numbers. To simplify matters, consider this example, using the twin prime numbers 41 and 43. Following the above strategy: $(41 + 58)^2 - 9 = (99)^2 - 9 = 9792 = 36 \times 272$, and the number $272 = 16 \times 17$, which are the consecutive numbers we originally sought.

Glorious Number Patterns

Admire these numbers and see what you can conclude regarding their patterns.

$12^2 = 144$	$122^2 = 14{,}884$
$1222^2 = 1{,}493{,}284$	$12{,}222^2 = 149{,}377{,}284$
$21^2 = 441$	$221^2 = 48{,}841$
$2221^2 = 4{,}932{,}841$	$22{,}221^2 = 493{,}772{,}841$
$122{,}222^2 = 14{,}938{,}217{,}284$	$1{,}222{,}222^2 = 1{,}493{,}826{,}617{,}284$
$222{,}221^2 = 49{,}382{,}172{,}841$	$2222221^2 = 4{,}938{,}266{,}172{,}841$

When we compare the powers of the numbers with the same amount of 2s, we notice that smaller number, such as the 1222^2, has 1 as the lead digit, as compared to the partner number 2221^2, where the digit 1 moves to the units place. Otherwise, the digits between these end digits are the same. This is a truly amazing relationship to be appreciated.

There is more to be appreciated with these number relationships, as we can see here:

$$21^2 - 12^2 = 297 = 9 \times 3$$
$$221^2 - 122^2 = 33957 = 99 \times 343$$
$$2221^2 - 1222^2 = 3439557 = 999 \times 3443$$
$$22{,}221^2 - 12{,}222^2 = 344{,}395{,}557 = 9999 \times 34{,}443$$
$$222{,}221^2 - 122{,}222^2 = 34{,}443{,}955{,}557 = 99{,}999 \times 344{,}443.$$

And this will continue on and on, preserving the relationship that should be very clear.

The Hidden Curiosities in Numbers with Repeating Digits

In Figure 1-38, we show numbers with repeating digits and wonder whether these numbers have any common property, such as prime numbers.

21 = 3 × 7
221 = 31 × 17
2 221 = PRIME NUMBER
22 221 = 3 × 3 × 3 × 823
222 221 = 359 × 619
2 222 221 = 19 × 116,959
22 222 221 = 3 × 7 × 373 × 2837
222 222 221 = 13 × 17,094,017
2 222 222 221 = 89 × 24,968,789
22 222 222 221 = 3 × 56,239 × 131,713
222 222 222 221 = 29 × 7,662,835,249
2 222 222 222 221 = 163 × 181 × 75,321,907
22 222 222 222 221 = 3 × 3 × 7 × 7 × 109 × 443 × 521 × 2003
222 222 222 222 221= 13 × 131 × 197,003 × 662,369
2 222 222 222 222 221 = 67 × 1229 × 131,321 × 205,507
22 222 222 222 222 221 = 3 × 23 × 211 × 239,851 × 6,363,769
222 222 222 222 222 221 = PRIME NUMBER
2 222 222 222 222 222 221 = 17 × 130,718,954,248,366,013

Figure 1-38

When we replace the 2s of the sequence in Figure 1-38 with 3s, we get a very different picture, as we can see in Figure 1-39. The first seven numbers are prime, and then we have to go way down to the seventeenth number before we get another prime number. However, the comparison to the previous sequence is clearly interesting and unusual.

31 = PRIME NUMBER
331 = PRIME NUMBER
3 331 = PRIME NUMBER
33 331 = PRIME NUMBER
333 331= PRIME NUMBER
3 333 331 = PRIME NUMBER
33 333 331 = PRIME NUMBER
333 333 331= 17 × 19,607,843
3 333 333 331 = 673 × 4,952,947
33 333 333 331 = 307 × 108,577,633
333 333 333 331 = 19 × 83 × 211,371,803
3 333 333 333 331 = 523 × 3049 × 2,090,353
33 333 333 333 331 = 607 × 1511 × 1997 × 18,199
333 333 333 333 331 = 181 × 1,841,620,626,151
3 333 333 333 333 331= 199 × 16,750,418,760,469
33 333 333 333 333 331 = 31 × 1499 × 717,324,094,199
333 333 333 333 333 331 = PRIME NUMBER
3 333 333 333 333 333 331 = 1009 × 3,303,600,925,008,259

Figure 1-39

Determining the Deleted Number

This is a characteristic with which you will impress your friends. Tell your friend to select a four-digit number without telling you what it is. Then tell him to find the sum of the digits of the number. Ask your friend to delete one of the four digits of the original number. With this new number, have your friend subtract the previous digit sum, then tell you the result of this calculation. You then take this number, add the digits, and then subtract this sum from the closest multiple of 9. The result will be the number that your friend deleted earlier.

To better follow this trick, we provide an example:

- Your friend selected the number 1836, whose digit sum is 18.
- He decided to delete 3 so that the remaining number is 186.
- From 186, he subtracts the original digit sum 18 to get 168, which he tells you.
- Now, you take the sum of the digits $1 + 6 + 8 = 15$, and then subtract it from the nearest multiple of 9, which is 18.
- The number he deleted is $18 - 15 = 3$, which will truly amaze them because it is in fact, the number he deleted.

Your task may then be to deal with your friend's amazement.

Creating a Strange Number

Begin by having your friend write the number 12,345,679 and ask which of these numbers she most adores. Say she selects the number **7**. Then have her multiply the original number by 63, whereupon she will be surprised that the result is 12,345,679 × 63 = **777,777,777,** which should make her very happy.

The trick is that the number 12,345,679 (which does not include the digit 8) when multiplied by 9 times any of the single digits 1–9 will result in a number that has nine of these digits. For example, 12,345,679 × (9 × 8) = 888,888,888. This is again an example of the secret wonders that can be found in mathematics.

Are These Right or Wrong?

In Figure 1-40, there are four challenges that seek a response. Are they true or false? These are truly amazing and unique relationships in mathematics. (Please don't look at the righthand column until you answer the questions.)

$12^2 = 144$, also $21^2 = 441$	Right
$13^2 = 169$, also $31^2 = 961$	Right
$\sqrt{5\frac{5}{24}} = 5\sqrt{\frac{5}{24}}$	Wrong
The lines joining the midpoints consecutively of any quadrilateral is a parallelogram.	Right

Figure 1-40

Guess the Number

Determine numbers n less than 20, where when you add x to the number you get a perfect square z, and when you subtract x from the number n you get $\sqrt{z}$.

For example, start with $n = 3$. When you add 1 and subtract $\sqrt{4}$, you get the same number as when you subtract 1 from n. Figure 1-41 presents a number of options fulfilling these requirements. A motivated reader might discover others.

	x	z	$\sqrt{z}$	$n-x$
3	1	4	2	2
10	6	16	4	4
15	10	25	5	5

Figure 1-41

Guess the Magic Number

Here is a challenge for the motivated reader. Find a three-digit number where if you subtract 7 from the number the result is a multiple of 7, and if you subtract 8 from the number the result is a multiple of 8, and furthermore, if you subtract 9 from the number the result is a multiple of 9.

The magic number needs to be a product of the three numbers 7, 8, and 9, which is 504. Therefore, $504 - 7 = 497 = 7 \times 71$, also $504 - 8 = 496 = 8 \times 62$, and $504 - 9 = 495 = 9 \times 55$.

To Find the Secret Number

The challenge here is to find the smallest number that when divided by 3 results in a remainder of 1, when divided by 4 results in a remainder of 2, when divided by 5 results in a remainder of 3, and when divided by 6 results in a remainder of 4.

In this case, the number always has a difference of 2 between the divisor and the remainder. Therefore 2+ the desired number must be a multiple of the divisors given. Since we are looking for the smallest such number, the lowest common multiple of 3, 4, 5, and 6 is 60. Thus, the number we seek is $60 - 2 = 58$. We can test to see if the number 58 in fact meets the required criteria. $\frac{58}{3} = 19$ remainder 1, $\frac{58}{4} = 14$ remainder 2, $\frac{58}{5} = 11$ remainder 3, and finally $\frac{58}{6} = 9$ remainder 4. This justifies that the number 58 meets the required criteria.

An Unusual Number Relationship

Consider any three-digit number. Then find all the two-digit numbers that use these three digits, add them, and divide the sum of those numbers by the sum of the digits of the original number. The amazing thing is that regardless of the original three-digit number selected, you will always end up with the number 22. Consider this example using the number 136. We form all the two-digit numbers using these 3 digits: 13, 16, 31, 36, 61, and 63. The sum of these numbers is 220. When we divide that by the sum of the digits of the original number 136, that is, $1 + 3 + 6 = 10$, we get $220 \div 10 = 22$.

Supposing you chose to add to the list of two-digit numbers the duplicates of the digits, namely, 11, 33, and 66, whose sum is 110. With these three additional numbers the total of the new set of nine numbers would always be 330 which would make the magic number after dividing by the sum of the digits of the original number 33. You may want to try this with another 3-digit number of your choice to verify this phenomenon.

Expected Digits

Here is a neat technique to develop a number which will contain the digits 1, 2, 4, 5, 7, and 8. All you need to do is follow these instructions:

- Select any number from 1 through 6
- Multiply the number by 9
- Multiply the number by 111
- Multiply the number by 1001
- Divide the last number by 7

The result will be a number which will contain the digits 1, 2, 4, 5, 7, and 8.

For example, consider the number 6 as the starting number:

$$6 \times 9 \times 111 \times 1001 \div 7 = 857{,}142.$$

As another example, consider the number 5:

$$5 \times 9 \times 111 \times 1001 \div 7 = 714{,}285.$$

An Unexpected Solution

Consider the sequence

$$(5 \cdot 7 \cdot 9) + (7 \cdot 9 \cdot 11) + (9 \cdot 11 \cdot 13) + (11 \cdot 13 \cdot 15) + \cdots +$$

(8th term), where we are challenged to find the eighth term and the sum of the first eight terms.

Following the scheme, we can determine that the eighth term is $(19 \cdot 21 \cdot 23)$. The process to find the sum of the end terms of this series is to write down the last term multiplied by the next highest factor, subtract the first term multiplied by the next lowest factor, and then divide by the product of the number of factors in each term +1 and the common difference of the factors in each term. Algebraically, this translates to the following:

$$\text{Sum} = \frac{19 \cdot 21 \cdot 23 \cdot 25 - 3 \cdot 5 \cdot 7 \cdot 9}{(3+1) \cdot 2} = \frac{229{,}425 - 945}{8} = 28{,}560.$$

Clever Foresight

The challenge here is to find the values of x and y in the following equation:

$$\left[3(230 + x)\right]^2 = 492y04.$$

Here is where we need a bit of foresight. The left side of the equation has a factor of $3^2 = 9$. Therefore, the sum of the digits on the right side must be a factor of 9, and so, $4 + 9 + 2 + y + 0 + 4 = 19 + y$. Therefore, $y = 8$.

We then solve the equation:

$$\sqrt{[3(230+x)]^2}=\sqrt{492804}$$
$$3(230+x)=702$$
$$x=4.$$

Cleverness is one of the best tools in solving mathematics problems. This is just one simple example of its usage.

Sum of Odd Numbers Generating Square Numbers—Graphically

One can generate square numbers by summing odd integers as follows: 1, 1 + 3 = 4, 1 + 3 + 5 = 9, 1 + 3 + 5 + 7 = 16, and so on. Representing this procedure graphically enables us to generate the algebraic version. In Figure 1-42, we show that in each case the previous square is augmented by twice the number of rows plus 1. Algebraically, each succeeding square can be represented as $n^2 + (2n+1)$, where n^2 is the size of the previous square.

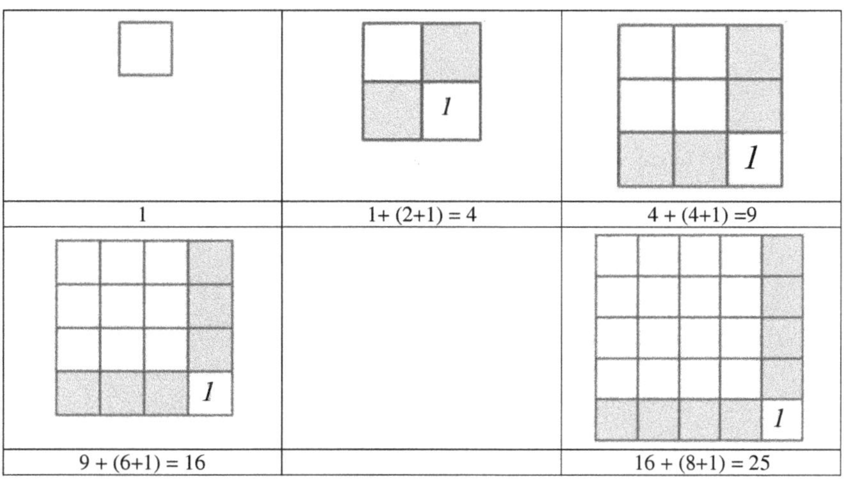

Figure 1-42

Find Hidden Patterns for Series Sums

Faced with summing a series, most folks would just plow right into the problem, using whatever means they learned to add the terms of the series and find its sum. This is drudgery, and very inelegant!

Let's look at a situation that lends itself to a few nifty alternatives. Consider the problem of finding the sum of the following series:

$$\frac{1}{12}+\frac{1}{23}+\frac{1}{34}+\cdots+\frac{1}{4950}.$$

One way to begin is to see if there is any pattern to be discovered. Here is one possibility:

$$\frac{1}{1\cdot 2}=\frac{1}{2}$$
$$\frac{1}{1\cdot 2}+\frac{1}{2\cdot 3}=\frac{2}{3}$$
$$\frac{1}{1\cdot 2}+\frac{1}{2\cdot 3}+\frac{1}{3\cdot 4}=\frac{3}{4}$$
$$\frac{1}{1\cdot 2}+\frac{1}{2\cdot 3}+\frac{1}{3\cdot 4}+\frac{1}{4\cdot 5}=\frac{4}{5}.$$

In this pattern, the sum of each series is an improper fraction formed by the factors of the last fraction in the series. Assume that the series that goes to $\frac{1}{4950}$. Therefore, we have the following sum:

$$\frac{1}{1\cdot 2}+\frac{1}{2\cdot 3}+\frac{1}{3\cdot 4}+\frac{1}{4\cdot 5}+\cdots+\frac{1}{49\cdot 50}=\frac{49}{50}.$$

Another pattern for this series can be obtained by representing each fraction in the series as a difference of two fractions in the following way:

$$\frac{1}{1 \cdot 2} = \frac{1}{1} - \frac{1}{2}$$
$$\frac{1}{2 \cdot 3} = \frac{1}{2} - \frac{1}{3}$$
$$\frac{1}{3 \cdot 4} = \frac{1}{3} - \frac{1}{4}$$
$$\vdots$$
$$\frac{1}{49 \cdot 50} = \frac{1}{49} - \frac{1}{50}.$$

By adding these equations, the left side gives us our sought-after sum, and on the right side all (except two) of the fractions drop out, leaving

$$\frac{1}{1} - \frac{1}{50} = \frac{49}{50}.$$

These very surprising techniques employing useful patterns get such reactions as, "Oh, I would never be able to do this on my own." This should be an unacceptable response, for "practice makes perfect!"

Are All Numbers Equal?

The title of this unit is clearly preposterous! But as you will see from the demonstration in Figure 1-43, such may be the case. We shall begin with the equation $\frac{x-1}{x-1} = 1$. Each succeeding row can be easily

	For $x = 1$
$\frac{x-1}{x-1} = 1$	$\frac{0}{0} = 1$
$\frac{x^2-1}{x-1} = x+1$	$\frac{0}{0} = 1+1 = 2$
$\frac{x^3-1}{x-1} = x^2+x+1$	$\frac{0}{0} = 1+1+1 = 3$
$\frac{x^4-1}{x-1} = x^3+x^2+x+1$	$\frac{0}{0} = 1+1+1+1 = 4$
$\vdots$	
$\frac{x^n-1}{x-1} = x^{n-1}+x^{n-2}+\cdots+x^2 + x + 1$	$\frac{0}{0} = 1+1+1+\cdots+1 = n$

Figure 1-43

justified with elementary algebra. There is nothing wrong with the algebra. Read this demonstration line-by-line and then draw your own conclusions.

When $x = 1$, then the numbers 1, 2, 3, 4, ..., n are each equal to $\frac{0}{0}$, which would make them all equal to each other. Of course, this cannot be true. For this reason, we define $\frac{0}{0}$ to be meaningless. Defining things to make them meaningful or consistent is what we do in mathematics to avoid ridiculous statements, as was the case here.

An Unsolved Challenge

In 1742, the Prussian mathematician Christian Goldbach (1690–1764) wrote a letter to the famous Swiss mathematician Leonhard Euler (1707–1783) posing the following problem, which to this day has yet to be solved.

Goldbach's Conjecture:
Every even number greater than 2 can be expressed as the sum of two prime numbers.

In Figure 1-44 we have a list of some even numbers and their corresponding sum of two prime numbers.

Even Numbers Greater Than 2	Sum of Two Prime Numbers
4	2 + 2
6	3 + 3
8	3 + 5
10	3 + 7
12	5 + 7
14	7 + 7
16	5 + 11
18	7 + 11
20	7 + 13
...	...
48	19 + 29
...	...
100	3 + 97

Figure 1-44

Can you find some more examples of this?

Goldbach's Second Conjecture:
Every odd number greater than 5 is the sum of three primes.

In Figure 1-45, we offer the first few odd numbers:

Odd Numbers Greater Than 5	Sum of Three Prime Numbers
7	2 + 2 + 3
9	3 + 3 + 3
11	3 + 3 + 5
13	3 + 5 + 5
15	5 + 5 + 5
17	5 + 5 + 7
19	5 + 7 + 7
21	7 + 7 + 7
...	...
51	3 + 17 + 31
...	...
77	5 + 5 + 67
...	...
101	5 + 7 + 89

Figure 1-45

These unsolved problems have tantalized many mathematicians over the centuries. Although no proof to justify these conjectures has yet been found, there is more evidence (with the help of computers) that these must be true, since no counterexample has been found. Interestingly, the efforts to solve these conjectures have led to some important discoveries in mathematics that might otherwise have remained hidden. For us, they provide sources for entertainment.

Howlers

In the early years of schooling, we learned to reduce fractions. There are specific ways to do it correctly. Some wise guy seems to have come up with a shorter way to reduce some fractions. Is he right?

He was asked to reduce the fraction $\frac{26}{65}$ and did it in the following way:

$$\frac{2\not{6}}{\not{6}5} = \frac{2}{5}.$$

That is, he just canceled out the 6s to get the right answer. Is this procedure correct? Can it be extended to other fractions? If so, then we were surely treated unfairly by our elementary school teachers who made us do much more work. Let's look at what was done here and if it can be generalized.

In his book, *Fallacies in Mathematics*, the Scottish mathematician E.A. Maxwell (1907–1987) refers to the following cancellations as howlers:

$$\frac{1\not{6}}{\not{6}4} = \frac{1}{4} \qquad \frac{2\not{6}}{\not{6}5} = \frac{2}{5}.$$

Perhaps when someone did the fraction reductions this way and still got the right answer, it could just make you howl.

Begin by reducing the following fractions to their lowest terms:

$$\frac{16}{64}, \quad \frac{19}{95}, \quad \frac{26}{65}, \quad \frac{49}{98}.$$

After you have reduced to lowest terms each of the fractions in the usual manner, one may ask why it couldn't have been done in the following way:

$$\frac{1\not{6}}{\not{6}4} = \frac{1}{4}$$

$$\frac{1\not{9}}{\not{9}5} = \frac{1}{5}$$

$$\frac{2\not{6}}{\not{6}5} = \frac{2}{5}$$

$$\frac{4\not{9}}{\not{9}8} = \frac{4}{8} = \frac{1}{2}.$$

At this point you may be somewhat amazed. Your first reaction is probably to ask if this can be done with any fraction composed of two-digit numbers, where the numerator and denominator share a digit.

Can you find another fraction (comprised of two-digit numbers) where this type of cancellation will work? You might cite

$$\frac{55}{55}=\frac{5}{5}=1$$

as an illustration of this type of cancellation. This will hold true for all two-digit multiples of eleven. Our concern will be only with proper fractions (i.e., whose value is less than one).

For those readers with a good working knowledge of elementary algebra, we can "explain" this situation. That is, why are the four fractions above the *only* ones (composed of two-digit numbers) where this type of cancellation will hold true?

Consider the fraction

$$\frac{10x+a}{10a+y}.$$

Canceling the a's in the above four fractions made the fractions equal to $\frac{x}{y}$. Therefore,

$$\frac{10x+a}{10a+y}=\frac{x}{y}.$$

This yields:

$$\begin{aligned} y(10x+a)&=x(10a+y)\\ 10xy+ay&=10ax+xy\\ 9xy+ay&=10ax. \end{aligned}$$

And so,

$$y=\frac{10ax}{9x+a}.$$

At this point we shall inspect this equation. It is necessary that x, y, and a are integers since they were digits in the numerator and denominator of a fraction. It is now our task to find the values of a and x for which y will also be integral.

To avoid a lot of algebraic manipulation, you will want to set up a chart that will generate values of y from

$$y=\frac{10ax}{9x+a}.$$

Remember that x, y, and a must be single-digit integers. Figure 1-46 shows a portion of the table you will be constructing. Notice that the cases where $x = a$ are excluded since $\frac{x}{a} = 1$.

x\a	1	2	3	4	5	6	…	9
1		$\frac{20}{11}$	$\frac{30}{12}$	$\frac{40}{13}$	$\frac{50}{14}$	$\frac{60}{15} = 4$		$\frac{90}{18} = 5$
2	$\frac{20}{19}$		$\frac{60}{21}$	$\frac{80}{22}$	$\frac{100}{23}$	$\frac{120}{24} = 5$		
3	$\frac{30}{28}$	$\frac{60}{29}$		$\frac{120}{31}$	$\frac{150}{32}$	$\frac{180}{33}$		
4								$\frac{360}{45} = 8$
⋮								
9								

Figure 1-46

The portion of the chart pictured above already generated two of the four integral values of y: that is, when $x = 1$ and $a = 6$, then $y = 4$; and when $x = 2$ and $a = 6$, then $y = 5$. These values yield the fractions $\frac{16}{64}$ and $\frac{26}{65}$, respectively. The remaining two integral values of y will be obtained when $x = 1$ and $a = 9$, yielding $y = 5$, and when $x = 4$ and $a = 9$, yielding $y = 8$. These yield the fractions $\frac{19}{95}$ and $\frac{49}{98}$, respectively. This should convince you that there are only four such fractions composed of two-digit numbers.

You may now wonder if this strange type of cancellation holds true for fractions composed of numerators and denominators of more than two digits. Try this type of cancellation with $\frac{499}{998}$. By crossing out the 9s, you should find that

$$\frac{499}{998} = \frac{4}{8} = \frac{1}{2}.$$

A pattern is now emerging, and you may realize that

$$\frac{49}{98}=\frac{499}{998}=\frac{4999}{9998}=\frac{49999}{99998}=\ldots.$$

$$\frac{16}{64}=\frac{166}{664}=\frac{1666}{6664}=\frac{16666}{66664}=\frac{166666}{666664}=\ldots.$$

$$\frac{19}{95}=\frac{199}{995}=\frac{1999}{9995}=\frac{19999}{99995}=\frac{199999}{999995}=\ldots.$$

$$\frac{26}{65}=\frac{266}{665}=\frac{2666}{6665}=\frac{26666}{66665}=\frac{266666}{666665}=\ldots.$$

Enthusiastic readers may wish to justify these extensions of the original howlers. Readers who, at this point, have a further desire to seek out additional fractions that permit this strange cancellation should consider the following fractions. They should verify the legitimacy of this strange cancellation and then set out to discover more such fractions.

$$\frac{3\not{3}2}{8\not{3}0}=\frac{32}{80}=\frac{2}{5}$$

$$\frac{3\not{8}5}{8\not{8}0}=\frac{35}{80}=\frac{7}{16}$$

$$\frac{1\not{3}8}{\not{3}45}=\frac{18}{45}=\frac{2}{5}$$

$$\frac{2\not{7}5}{7\not{7}0}=\frac{25}{70}=\frac{5}{14}$$

$$\frac{1\not{6}\not{3}}{\not{3}2\not{6}}=\frac{1}{2}.$$

Aside from providing an algebraic application, which can introduce a number of important topics in a motivational way, this topic can also provide some recreational activities. Here are some more of these "howlers."

$$\frac{4\cancel{84}}{\cancel{84}7}=\frac{4}{7}\qquad \frac{\cancel{54}5}{6\cancel{54}}=\frac{5}{6}\qquad \frac{\cancel{42}4}{7\cancel{42}}=\frac{4}{7}\qquad \frac{24\cancel{9}}{\cancel{9}96}=\frac{24}{96}=\frac{1}{4}$$

$$\frac{4\cancel{8484}}{\cancel{8484}7}=\frac{4}{7}\qquad \frac{\cancel{5454}5}{6\cancel{5454}}=\frac{5}{6}\qquad \frac{\cancel{4242}4}{7\cancel{4242}}=\frac{4}{7}$$

$$\frac{\cancel{324}3}{4\cancel{324}}=\frac{3}{4}\qquad \frac{\cancel{648}6}{8\cancel{648}}=\frac{6}{8}=\frac{3}{4}$$

$$\frac{14\cancel{714}}{\cancel{714}68}=\frac{14}{68}=\frac{7}{34}\qquad \frac{\cancel{87804}8}{9\cancel{87804}}=\frac{8}{9}$$

$$\frac{1\cancel{428571}}{\cancel{428571}3}=\frac{1}{3}\qquad \frac{2\cancel{857142}}{\cancel{857142}6}=\frac{2}{6}=\frac{1}{3}\qquad \frac{3\cancel{461538}}{\cancel{461538}4}=\frac{3}{4}$$

$$\frac{\cancel{76712328}7}{8\cancel{76712328}}=\frac{7}{8}\qquad \frac{\cancel{324324324}3}{4\cancel{324324324}}=\frac{3}{4}$$

$$\frac{\cancel{102564}1}{4\cancel{102564}}=\frac{1}{4}\qquad \frac{\cancel{324324}3}{4\cancel{324324}}=\frac{3}{4}\qquad \frac{4\cancel{571428}}{\cancel{571428}5}=\frac{4}{5}$$

$$\frac{4\cancel{848484}}{\cancel{848484}7}=\frac{4}{7}\qquad \frac{5\cancel{952380}}{\cancel{952380}8}=\frac{5}{8}\qquad \frac{\cancel{428571}4}{6\cancel{428571}}=\frac{4}{6}=\frac{2}{3}$$

$$\frac{\cancel{545454}5}{6\cancel{545454}}=\frac{5}{6}\qquad \frac{6\cancel{923076}}{\cancel{923076}8}=\frac{6}{8}=\frac{3}{4}\qquad \frac{\cancel{424242}4}{7\cancel{424242}}=\frac{4}{7}$$

$$\frac{\cancel{538461}5}{7\cancel{538461}}=\frac{5}{7}\qquad \frac{\cancel{205128}2}{8\cancel{205128}}=\frac{2}{8}=\frac{1}{4}\qquad \frac{\cancel{311688}3}{8\cancel{311688}}=\frac{3}{8}$$

$$\frac{\cancel{648648}6}{8\cancel{648648}}=\frac{6}{8}=\frac{3}{4}\qquad \frac{4\cancel{84848484}}{\cancel{84848484}7}=\frac{4}{7}$$

This discussion shows how elementary algebra can be used to investigate number theory, one that is also quite amusing. Mathematics continues to hold some hidden treasures.

Chapter 2

Arithmetic Novelties

Arithmetic has ever been the basis for applied mathematics in everyday life. Although in today's world people generally outsource their arithmetic to calculators and computers, arithmetic still has a flair to it if it is seen from a creative point of view. In this chapter, we will explore multiple shortcuts for performing arithmetic which avoid traditional calculations and obviate the need for a calculator. Some useful problem-solving techniques will also be presented in this chapter. Finally, we will look at some unusual calculation processes from other cultures. In some cases, these techniques may be too antiquated to be practical, but they can give you interesting insights into mathematics.

Multiplication by Divisors of Powers of 10

Multiplying by factors (or divisors) of powers of 10 is just a bit more involved than merely multiplying by powers of 10. However, in many cases, you can perform such functions mentally. Let's begin by considering multiplication by 25 (a factor of 100). Perhaps it is best that we begin with an example: 16×25. Since $25 = \frac{100}{4}$, we can write this as $16 \times 25 = 16 \times \frac{100}{4} = \frac{16}{4} \times 100 = 4 \times 100 = 400$. Rather simple! All we did was divide the multiplicand by 4 and then multiply by 100, which means appending two 0s to the number. However, a little further practice can be helpful, so here are some more examples:

38×25 can be written as

$$38 \times \frac{100}{4} = \frac{38}{4} \times 100 = \frac{19}{2} \times 100 = 9.5 \times 100 = 950.$$

1.7×25 can be written as

$$\frac{17}{10} \times \frac{100}{4} = \frac{17}{4} \times \frac{100}{10} = 4.25 \times 10 = 42.5.$$

In the previous example, we converted

$$\frac{17}{4} = \frac{16}{4} + \frac{1}{4} = 4 + 0.25 = 4.25.$$

In some cases, these tricks are more entertaining than useful, but they do give some interesting insights into why we learned various arithmetic procedures in school.

In an analogous fashion, we can multiply by 125, since $125 = \frac{1000}{8}$. Here are some examples of multiplication by 125—mentally!

$$32 \times 125 = 32 \times \frac{1000}{8} = \frac{32}{8} \times 1000 = 4 \times 1000 = 4000$$

$$78 \times 125 = 78 \times \frac{1000}{8} = \frac{78}{8} \times 1000 = \frac{39}{4} \times 1000 = 9.75 \times 1000 = 9750$$

$$3.4 \times 125 = 3.4 \times \frac{1000}{8} = \frac{3.4}{8} \times 1000 = \frac{1.7}{7} \times 1000 = 0.425 \times 1000 = 425.$$

When multiplying by 50, you can use $50 = \frac{100}{2}$, or when multiplying by 20, you can use $20 = \frac{100}{5}$. Practice with these special numbers will clearly help you to do various calculations.

A Simple Trick for Dividing by 5

In many cases, you can divide by 5 mentally. However, to dramatize the effectiveness of this technique, we will demonstrate it with a larger number. Consider dividing 5678 by 5. The process is as follows: double the number 5678 to get 11,356, and then simply move the decimal point one place to the left to get 1135.6. It's that simple!

Multiplying by 11—A Useful Technique

The fact that the number 11 is one greater than the base 10 enables a very useful multiplication technique: *To multiply a two-digit number by 11, just add the two digits of the number multiplied by 11 and place this sum between the two digits.*

For example, suppose you have to multiply 45 by 11. According to this technique, add $4 + 5 = 9$ and place the 9 between the 4 and 5 to get 495. This can get a bit more difficult when the sum of the two digits you added results in a two-digit number. If the sum of the two digits is greater than 9, then we place the units digit between the two digits of the number being multiplied by 11 and "carry" the tens digit to be added to the hundreds digit of the multiplicand. Consider as an example 78×11. As before, we add the two digits $7 + 8 = 15$. We place the 5 between the 7 and 8, and add the 1 to the 7 to get [7+1][5][8], or 858.

The technique also holds when a number with more than two digits is multiplied by 11. For example, we shall apply this technique to the multidigit number 12,345, which we will multiply by 11. As before, we begin at the right-side digit and add every pair of digits going to the left: 1[1+2][2+3][3+4][4+5]5 = 135,795. If the sum of two digits is greater than 9, then use the procedure described above: place the units digit appropriately and carry the tens digit. As an example, multiply 456,789 by 11.

We carry out the process step by step:

4[4+5][5+6][6+7][7+8][8+9]9
4[4+5][5+6][6+7][7+8][17]9
4[4+5][5+6][6+7][7+8+*1*][7]9
4[4+5][5+6][6+7][16][7]9
4[4+5][5+6][6+7+*1*][6][7]9
4[4+5][5+6][14][6][7]9
4[4+5][5+6+*1*][4][6][7]9
4[4+5][12][4][6][7]9
4[4+5+*1*][2][4][6][7]9
4[10][2][4][6][7]9
[4+*1*][0][2][4][6][7]9
[5][0][2][4][6][7]9
5,024,679.

Obviously, the trick to multiply by 11 is most impressive with small numbers, where the process can be done mentally and be very impressive to an audience. However, it is also interesting to see how this can be extended to larger numbers, as shown above.

A Clever Technique to Determine Divisibility by 11

Supposing you wish to know if a particular number is divisible by 11. Certainly, one can use a calculator and determine if the answer is an integer. The number 11 lends itself to a very clever technique for determining whether a given number is divisible by 11. The technique is as follows:

- Take the sum of the odd-placed digits of the original number.
- Take the sum of the remaining digits.
- If the difference between these two sums is a multiple of 11,
- Then the original number was a multiple of 11.

As an example, let's determine if the number 1,357,531 is divisible by 11. The sum of the odd-placed digits is $1 + 5 + 5 + 1 = 12$. The sum of the remaining digits is $3 + 7 + 3 = 13$. The difference of these two sums is $13 - 12 = 1$, which is not a multiple of 11. Therefore, the original number, 1,357,531, is *not* a multiple of 11. On the other hand, consider 135,795. Take the sum of the odd placed digits $1 + 5 + 9 = 15$, and the sum of the remaining digits, $3 + 7 + 5 = 15$. The difference of the sums is $15 - 15 = 0$, which is a multiple of 11. Therefore, the original number 135,795 is a multiple of 11.

For the interested reader, here is a brief discussion about why this rule works.

Consider the number $\overline{abcde}$, where $a, \ldots, e \in \{0, 1, 2, 3, \ldots, 9\}$ and $a \neq 0$, whose value can be expressed as

$$\begin{aligned}
N &= 10^4 a + 10^3 b + 10^2 c + 10d + e \\
&= (11-1)^4 a + (11-1)^3 b + (11-1)^2 c + (11-1)d + e \\
&= [11M + (-1)^4]a + [11M + (-1)^3]b + [11M + (-1)^2]c + [11 + (-1)]d + e \\
&= 11M[a + b + c + d] + a - b + c - d + e,
\end{aligned}$$

which implies that divisibility by 11 of N depends on the divisibility of:

$$a - b + c - d + e = (a + c + e) - (b + d),$$

the difference of the sums of the alternate digits.

Note: $11M$ denotes a multiple of 11.

Another Clever Trick to Determine Divisibility by 11

There is another way to determine if a number is divisible by 11. Take a number and break it up into pairs of adjacent digits. If the number consists of an odd number of digits, place a zero on either end of the number. Find the sum of these pairs of two-digit numbers. If the resulting number is divisible by 11, then the original number is divisible by 11.

Suppose we would like to determine if the number 786,432,614 is divisible by 11. We partition this long number into pairs of consecutive digits as follows: 07 + 86 + 43 + 26 + 14 = 176. We continue this process: 01 + 76 = 77, which is a multiple of 11. Therefore, the original number is divisible by 11.

How to Determine Divisibility by 3 or 9

To determine if a given number is divisible by 3 or 9, one simply needs to find the sum of the digits of this number, and if that digit sum is divisible by 3 or 9, then the original number is divisible by 3 or 9.

For example, suppose we would like to determine if the number 36,180,225 is divisible by 3 or 9. We take the sum of the digits: 3 + 6 + 1 + 8 + 0 + 2 + 2 + 5 = 27. Although we know that the number 27 is divisible by 3 and 9, we could still take this a step further by adding 2 + 7 = 9, and then conclude that the original number 36,180,225 is divisible by 9 and 3.

As another example, determine if the number 854,724 is divisible by 3 and 9. Take the sum of the digits: 8 + 5 + 4 + 7 + 2 + 4 = 30, which is divisible by 3 but not by 9. Therefore, the number 854,724 is divisible by 3 and not by 9.

Considering Divisibility by Powers of 2 and 5

Previously, we presented a nifty little technique for determining if a number is divisible by 3 or by 9. Most adults can determine when a number is divisible by 2 or by 5 simply by looking at the units digit of the number. That is, if the last digit is an even number, then the number will be divisible by 2. Incidentally, if the original number's last two digits form a multiple of 4, then the number itself is divisible by 4. Also, if the number formed by the last three digits is divisible by 8, then the original number is divisible by 8. You can extend this rule to check for divisibility by higher powers of 2 as well.

Similarly for 5, if the last digit of the number being inspected for divisibility is either a 0 or 5, then the number itself will be divisible by 5. If the number formed by the last two digits is divisible by 25, then the original number itself is divisible by 25. This is analogous to the rule for powers of 2. Have you guessed what the relationship here is between powers of 2 and 5? Yes, they are the factors of 10, the basis of our decimal number system.

Divisibility by Prime Numbers

Divisibility rules provide an interesting "window" into the nature of numbers and their properties and furnish a better appreciation for mathematics. For this reason (among others), the topic of divisibility still finds a place on the mathematics-learning spectrum.

Establishing rules for checking divisibility by prime numbers has always been most perplexing. The tricks for some of the divisibility rules for prime numbers are almost as cumbersome as an actual division algorithm, yet they are fun, and, believe it or not, they can be useful.

As an example, if you raise any prime (other than 2 or 3) to the fourth power and then subtract 1, the result will always have a factor of 48. If we wish to test a number to see if it is a prime, we take it to the fourth power and subtract 1. If the resulting number has a factor of 48, then the original number is a prime number. For example, let's consider the prime number 13. Following this technique, we get $13^4 - 1 = 28{,}561 - 1 = 28{,}560 = 48 \times 595$. Now let's test the number 27 to see if it is a prime number. $27^4 - 1 = 531{,}441 - 1 = 531{,}440$, and $\frac{531440}{48} = 11{,}071.66$. Therefore, since 48 is not a factor of 531,440, the number 27 is not a prime number—which we obviously already knew.

Let us begin by considering the technique for determining divisibility by 7, and then inspect it to see how this can be generalized for other prime numbers.

The Rule for Divisibility by 7:

Delete the last digit from the given number and then subtract twice this deleted digit from the remaining number. If the result is divisible by 7, the original number is divisible by 7. This process may be repeated if the result is too large for simple inspection of divisibility of 7.

To best understand this divisibility rule, suppose we want to test the number 876,547 for divisibility by 7. Begin with 876,547 and delete its units digit, 7, and subtract its double, 14, from the remaining number: $87{,}654 - 14 = 87{,}640$. Since we cannot yet visually inspect the resulting number for divisibility by 7, we continue the process. Continue with the resulting number 87,640 and delete its units digit, 0, and subtract its double, still 0, from the remaining number: $8764 - 0 = 8764$. Since the resulting number, 8,764, is not obviously divisible by 7, we continue the process. Continue with the resulting number 8,764 and delete its units digit, 4, and subtract its double, 8, from the remaining number: $876 - 8 = 868$. Since we still cannot visually inspect 868 for divisibility by 7, we continue the process. Continuing with the resulting number 868, we delete its units digit, 8, and subtract its double, 16, from the remaining number: $86 - 16 = 70$, which is divisible by 7. Therefore, the number 876,547 is divisible by 7.

Before continuing with our discussion of divisibility by prime numbers, you ought to practice this technique with a few randomly selected numbers and then check your results with a calculator.

Now for the beauty of mathematics! Why does this rather strange procedure actually work? The wonderful thing about mathematics is that (for the most part) it does not do things that we cannot justify.[1] This technique will make sense to you after you see what is happening with this procedure.

To justify the technique of determining divisibility by 7, consider the various possible terminal digits that you are "dropping" and the corresponding subtraction that is actually being done by dropping the last digit. Figure 2-1 shows how doubling the units digit and subtracting that from the number remaining when the units digit has been deleted results in a multiple of 7. That is, we have taken "bundles of 7" away from the original number. Therefore, if the remaining number is divisible by 7, then so is the original number, because you have separated the original number into parts, each of which is divisible by 7. Therefore, the entire number must be divisible by 7.

Terminal Digit	Number Subtracted from Original	Terminal Digit	Number Subtracted from Original
1	$20 + 1 = 21 = 3 \times 7$	5	$100 + 5 = 105 = 15 \times 7$
2	$40 + 2 = 42 = 6 \times 7$	6	$120 + 6 = 126 = 18 \times 7$
3	$60 + 3 = 63 = 9 \times 7$	7	$140 + 7 = 147 = 21 \times 7$
4	$80 + 4 = 84 = 12 \times 7$	8	$160 + 8 = 168 = 24 \times 7$
		9	$180 + 9 = 189 = 27 \times 7$

Figure 2-1

The Rule for Divisibility by 13:

This is the similar to the rule for testing divisibility by 7, except that instead of subtracting twice the deleted digit, we subtract 9 times the deleted digit each time.

For example, use the technique to determine divisibility by 13 for the number 5616. Begin with 5616 and delete its units digit, 6, and subtract its product by 9, 54, from the remaining number: $561 - 54 = 507$. Since we cannot visually inspect the resulting number

---

[1] There are a few phenomena in mathematics that have not yet found an acceptable justification (or proof), but that doesn't mean we won't find one in the future. It took us 350 years to justify Fermat's conjecture! It was done by Dr. Andrew Wiles several years ago.

for divisibility by 13, we continue the process. Continue with the resulting number 507, delete its units digit and subtract 9 times this digit from the remaining number to get 50 – 63 = –13, which is divisible by 13. Therefore, the original number is divisible by 13.

To determine the "multiplier," 9, we sought the smallest multiple of 13 that ends in a 1. That was 91, where the tens digit is *9 times* the units digit. Once again, consider the various possible terminal digits and the corresponding subtractions shown in Figure 2-2.

Terminal Digit	Number Subtracted from Original	Terminal Digit	Number Subtracted from Original
1	90 + 1 = 91 = 7 × 13	5	450 + 5 = 455 = 35 × 13
2	180 + 2 = 182 = 14 × 13	6	540 + 6 = 546 = 42 × 13
3	270 + 3 = 273 = 21 × 13	7	630 + 7 = 637 = 49 × 13
4	360 + 4 = 364 = 28 × 13	8	720 + 8 = 728 = 56 × 13
		9	810 + 9 = 819 = 63 × 13

Figure 2-2

In each case, a multiple of 13 is being subtracted one or more times from the original number. Hence, if the remaining number is divisible by 13, then the original number is divisible by 13.

The Rule for Divisibility by 17:

Delete the units digit and subtract 5 times the deleted digit each time from the remaining number until you reach a number small enough to determine its divisibility by 17.

We justify the rule for divisibility by 17 as we did the rules for 7 and 13. Each step of the procedure subtracts a "bunch of 17s" from the original number until we reduce the number to a manageable size and can make a visual inspection of divisibility by 17. Employing the technique for divisibility by 17 to the number 7871 begins with dropping the units digit, 1, and multiplying it by 5, which we subtract from 787 – 5 = 782. Continuing this process, we drop the 2, and multiply it by 5, subtract that from 78, and we get 78 – 10 = 68.

Supposing we want to test if 68 is divisible by 17, we could continue the process so that 6 – 40 = –34, which is divisible by 17.

The patterns developed in the preceding three divisibility rules (for 7, 13, and 17) should lead you to develop similar rules for testing divisibility by larger primes. Figure 2-3 presents the "multipliers" of the deleted digits for various tricks to check divisibility by primes.

To test divisibility by	7	11	13	17	19	23	29	31	37	41	43	47
Multiplier	2	1	9	5	17	16	26	3	11	4	30	14

Figure 2-3

With the experience of these divisibility rules for prime numbers, a motivated reader should be able to develop divisibility rules for other prime numbers, such as 19, where the multiplier would be 17.

A Curious Technique to Check Addition

Here we present a very famous technique for checking the addition of several numbers called "casting out 9s." To check an addition problem, all we need to do is to find the sum of the digits of each of the addends, add the digit sums, and compare the result to the sum of the digits of the calculated sum. If the sums are the same, then the answer is correct. We can show this with the following addition computation:

$$\begin{array}{r} 13579 \\ +86327 \\ \hline 99906 \end{array}$$

We take the digit-sum of each of the numbers above, continuing this process of digit sums until a single digit is reached. The digit-sum of the addends must be equal to the digit-sum of the final sum, as we can see below.

13579 — digit-sum: 25 — digit-sum: 7

+ 86327 — digit-sum: 26 — digit-sum: 8

sum of digit-sums: 7 + 8 = 15 — digit-sum: 6

99906 — digit-sum: 33 — digit-sum: 6

Since the sum of the digit-sums of the two addends (eventually) is 6, and it matches the digit-sum of the calculated sum, also 6, the addition could be correct.

Thou Shalt Not Divide by Zero

But why is division by zero not permissible? Mathematicians pride ourselves on the order, beauty, and consistency of everything in the realm of mathematics. When something arises that could spoil this order, we simply *define* it to suit our needs. This is precisely what happens with division by zero. One gets a much better insight into the nature of mathematics by explaining why these "rules" are set forth. To justify this rule, consider the following. We can speculate what the quotient $\frac{n}{0}$ is equal to by assuming it is equal to q; that is, $\frac{n}{0}=q$. If we check to see if $0 \cdot q = n$, we know it would be absurd, since $0 \cdot q = 0$. In other words, there is no number q that can take on the quotient to this division. For that reason, we define division by zero to be invalid.

A more convincing case for defining away division by zero is to show how it can lead to a contradiction of an accepted fact, namely, that $1 \neq 2$. Were division by zero acceptable, then $1 = 2$, which is clearly an absurdity! Consider the "proof" that $1 = 2$:

Let $a = b$	
Then $a^2 = ab$	[multiplying both sides by a]
$a^2 - b^2 = ab - b^2$	[subtracting b^2 from both sides]
$(a-b)(a+b) = b(a-b)$	[factoring]
$a+b = b$	[dividing by $(a-b)$]
$2b = b$	[replace a by b]
$2 = 1$	[divide both sides by b]

In the step where we divided by $(a - b)$, we actually divided by zero, because $a = b$, so that $a - b = 0$. That ultimately led us to an absurd result, leaving us with no option other than to prohibit division by zero. By taking the time to explain this rule about division by zero you will have a much better appreciation for the nature of mathematics. If you wish, there are other such absurdities that can be used to entertain people besides just showing that you can "prove" $1 = 2$.

Cleverly Combining Successive Percentages

It is typically difficult to precisely combine successive percentages—discounts and increases. Here is a technique to turn this nemesis of students into a delightfully simple arithmetic algorithm that affords lots of useful applications and provides new insight into successive percentage problems. This not-very-well-known procedure should enchant you. Let's begin by considering the following problem:

> *Wanting to buy a coat, Barbara is faced with a dilemma. Two competing stores next to each other carry the same brand of coat with the same list price, but with two different discount offers. Store A offers a 10% discount year-round on all its goods, but on this particular day offers an additional 20% on top of their already discounted price. Store B simply offers a discount of 30% on that day in order to stay competitive. Which store offers the best price and how many percentage points difference is there between the two options available to Barbara?*

At first glance, you may assume there is no difference in price, since $10 + 20 = 30$, yielding the same discount in both cases. Yet with a little more thought you will realize that this is not correct, since store A calculates a 10% discount on the original list price and a 20% discount calculated on the lower price, while store B calculates the entire 30% discount on the original price. Now, the question to be answered is, what percentage difference is there between the discounts in store A and store B?

One solution might be to assume the cost of the coat to be \$100, calculate the 10% discount yielding a \$90 price, and take an additional 20% off the \$90 price (or \$18), thus bringing the price down to \$72. In store B, the 30% discount on \$100 would bring the price down to \$70, giving a discount difference of \$2, which in this case is 2%. This procedure, although correct and not too difficult, is a bit cumbersome and does not always allow a full insight into the situation.

An interesting and quite unusual procedure is provided for entertainment and a fresh look into this problem. We offer here a method for obtaining a single percentage discount (or increase) equivalent to two (or more) successive discounts (or increases).

(1) Change each of the percents involved into decimal form:

$$.20 \text{ and } .10$$

(2) Subtract each of these decimals from 1.00:

$$.80 \text{ and } .90 \text{ (for an increase, add to 1.00)}$$

(3) Multiply these differences:

$$(.80)(.90) = .72$$

(4) Subtract this number (i.e., .72) from 1.00:

$$1.00 - .72 = .28, \text{ which represents the combined discount}$$

(If the result of step 3 is greater than 1.00, subtract 1.00 from it to obtain the percent of increase.)

When we convert .28 back to percent form, we obtain 28%, the equivalent of successive discounts of 20% and 10%.

This combined percentage of 28% differs from 30% by 2%.

Following the same procedure, you can also combine more than two successive discounts. In addition, successive increases, combined or not combined with a discount, can also be accommodated in this procedure by adding the decimal equivalent of the increase to 1.00 (whereas the discount was subtracted from 1.00) and then continuing the procedure in the same way. If the end result comes out greater than 1.00, then this reflects an overall increase rather than a discount (as found in the above problem).

This procedure not only streamlines a typically cumbersome situation but also provides some insight into percentages. For example, the question "Is it advantageous to the buyer in the above problem to receive a 20% discount and then a 10% discount, or the reverse, a 10% discount and then a 20% discount?" The answer to this question is not intuitively obvious. Yet, since the procedure just presented shows that the calculation is merely multiplication, a commutative operation, we find immediately that there is no difference between the two.

So here you have a delightful algorithm for combining successive discounts, increases, or combinations of both. Not only is this algorithm useful, but it also gives you some newfound power in dealing with percentages when using a calculator is not appropriate.

What causes some confusion is if a store offers a 10% discount one day and for some unknown reason decides to raise the price by 10%. One might assume that the new price is identical to the original price. However, when we use this new trick procedure, we find that the discount yields .90, and the increase yields 1.10, and the product of the two is .99, which unexpectedly indicates a discount of 1%.

The Rule of 72

Here we present a very useful method for determining how long it takes to double your money in the bank when it is compounded daily at a given annual rate. This is clearly good to know, but it is the unusualness of this rule that allows us to exhibit it here. We shall call it the "Rule of 72," as it is based on this number, as you will soon see.

The "Rule of 72" states that, roughly speaking, *money will double in* $\frac{72}{r}$ *years when it is invested at an annual compounded interest rate of* $r\%$. So, for example, if we invest money at an 8% compounded annual interest rate, it will double its value in $\frac{72}{8}=9$ years. Similarly, if we leave our money in the bank at a compounded rate of 6%, it would take 12 years for this sum to double its value.

The interested reader might want to understand why this is so, and how accurate the rule really is. The following discussion will explain. To investigate why, or if, this really works, we consider the compound interest formula: $A=p(1+\frac{r}{100})^n$, where A is the resulting amount of money, P is the principal invested for n interest periods at $r\%$ annually. We need to investigate what happens when $A = 2P$. The above equation then becomes

$$2=\left(1+\frac{r}{100}\right)^n. \tag{1}$$

It then follows that

$$n=\frac{\log 2}{\log\left(1+\frac{r}{100}\right)}. \tag{2}$$

In Figure 2-4, we have a table of values from the above equation, generated with the help of a scientific calculator.

r	*n*	*nr*
1	69.66071689	69.66071689
3	23.44977225	70.34931675
5	14.20669908	71.03349541
7	10.24476835	71.71337846
9	8.043231727	72.38908554
11	6.641884618	73.0607308
13	5.671417169	73.72842319
15	4.959484455	74.39226682

Figure 2-4

The arithmetic mean (the usual average) of the *nr* values is 72.04092314, which is quite close to 72, and so our "Rule of 72" seems to be an accurate estimate for doubling money at an annual interest rate of *r*% for *n* interest periods.

An ambitious reader, or one with a very strong mathematics background, might try to determine a "rule" for tripling and quadrupling money, similar to the way we dealt with the doubling of money. The above equation (2) for *k*-tupling would be

$$n=\frac{\log k}{\log\left(1+\frac{r}{100}\right)},$$

which for $r = 8$ gives the value for $n = 29.91884022$ (log *k*). Thus $nr = 239.3507218 \log k$, which for $k = 3$ (the tripling effect) gives us

nr = 114.1993167. We could then say that for tripling money we would have a "rule of 114."

However far this topic is explored, the important issue here is that the "Rule of 72" can be a handy tool to have at your disposal.

Arithmetic Challenges

Figure 2-5 shows arithmetic calculations where digits are missing. The challenge is to insert the missing digits that make the calculations correct.

Addition	Subtraction
□22□ 1□□1 3489	4□□2 □35□ 121
113 6□4 14□ □26 □410	6□35 □82□ 4□7

Figure 2-5

The correct arithmetic calculations are provided in Figure 2-6.

Addition Solutions		Subtraction Solutions	
2228 1261 3489	113 624 147 526 1410	4472 4351 121	6235 5828 407

Figure 2-6

Consider this a warmup for the last challenge of this chapter.

Comparing Radicals

The task here is to determine which is greater: $\sqrt{10}+\sqrt{17}$ or $\sqrt{53}$. We need to combine the two radicals to make a comparison. One possibility is to take the sum and square it, which is $\left(\sqrt{10}+\sqrt{17}\right)^2 = 10+2\sqrt{170}+17 = 27+2\sqrt{170}$. We must represent the number 53 in a form involving a square that can compare to $\sqrt{170}$.

This could be $\sqrt{169}=13$. Therefore, we have $53=27+2\sqrt{169}$, which is clearly less than $27+2\sqrt{170}$. So, we conclude that $\sqrt{53}<\sqrt{10}+\sqrt{17}$.

Comparing Fractions

There is an automatic way to compare the size of two fractions by placing them side-by-side and taking the cross products, as shown below. Suppose we want to compare the size of the two fractions $\frac{7}{9}$ and $\frac{9}{11}$. We take the two cross products: $7\times 11=77$ and $9\times 9=81$. Since $81>77$, the fraction whose numerator produced the larger cross product is the larger fraction, that is, $\frac{9}{11}>\frac{7}{9}$.

Find the Improper Fraction

Suppose one-third of a certain fraction plus three times that fraction is equal to that fraction multiplied by itself. The challenge is to find the fraction.

We can begin by noting the desired fraction as $\frac{x}{y}$. By following the description, we get

$$\frac{x}{y}\cdot\frac{1}{3}+3\cdot\frac{x}{y}=\frac{x}{y}\cdot\frac{x}{y},$$

which we can simplify as

$$\frac{x}{3y}+\frac{3x}{y}=\frac{x}{3y}+\frac{9x}{3y}=\frac{10x}{3y}=\frac{x^2}{y^2}.$$

Thus,

$$\frac{x}{y}=\frac{10}{3}.$$

Getting into an Endless Loop

The decimal system that we are accustomed to presents some unusual phenomena. There isn't much you can do about it other than marvel at the outcome. It is one of the arithmetic oddities in mathematics that is still searching for a justification. This is not something we can prove true for all cases, yet no numbers have been found for which it won't work. That, in itself, suffices to establish that it is apparently always true. You may wish to use a calculator, unless you enjoy subtraction.

Follow the steps below

1. Begin by selecting a four-digit number (except one that has all digits the same).
2. Rearrange the digits of the number so that they form the largest number possible.
3. Then rearrange the digits of the number so that they form the smallest number possible.
4. Subtract these two numbers (obviously, the smaller from the larger).
5. Take this difference and continue the process over and over and over, until you notice something unusual happening. Don't give up before the strange occurrence appears: when you reach 6174, you will find yourself in an endless loop.

When you have reached the loop, remember that you began with an arbitrarily selected number. Isn't this an astonishing result? To convince yourself that this will occur again, try selecting another four-digit number and repeat the process.

Here is an example of how this works with the arbitrarily selected starting number 3203.

We select the number 3203.

The *largest* number formed with these digits is 3320.

The *smallest* number formed with these digits is 0233.

The difference is 3087.

Now using this number, 3087, we continue the process.

The largest number formed with these digits is 8730.

The smallest number formed with these digits is 0378.

The difference is 8352.

Again, we repeat the process.

The largest number formed with these digits is 8532.

The smallest number formed with these digits is 2358.

The difference is **6174**.

The largest number formed with these digits is 7641.

The smallest number formed with these digits is 1467.

The difference is 6174.

And so the loop is formed, since you keep on getting 6174. Remember, all this began with an arbitrarily selected number, and it will always end up with the number 6174, which then gets you into an endless loop where we continuously return to 6174.

A Strange Division

When you divide a number consisting of the same digits by the sum of the digits, you will find that the answer is always the same for other repunits of the same quantity of digits. For example, any three-digit number consisting of the same digits will produce 37 through this algorithm, such as $555 \div 15 = 37$ and $888 \div 24 = 37$. Four-digit numbers consisting of the same digits divided by the sum of the digits will always result in 277.75, such as $7777 \div 28 = 277.75$. With a two-digit number, the quotient will always be 5.5, as we see with $66 \div 12 = 5.5$. This is truly amazing and should be appreciated.

An Unexpected Division Result

If you have a six-digit number where the first three digits are the same as the second three digits and divide this number by 7, 11, and 13 in any order, you will always result in the number which has been repeated twice in the original six-digit number.

For example, consider the six-digit number 578,578. When we divide this number by 7, we get 82,654. When we divide 82,654 by 11, we get 7514. Finally, when we divide 7514 by 13, we get 578, which is the three-digit number repeated in the original number. Remember, the division by 7, 11, and 13 could have been done in any order.

An Interesting Algebraic Phenomenon

If you reverse the coefficients of any polynomial equation, the resulting equation will have roots that are the reciprocals of the original roots. That is, for example $ax^4 + bx^3 + cx^2 + dx + e = 0$ has roots $\alpha, \beta, \gamma, \delta$. Then the equation with the coefficients reversed will have reciprocal roots:

$$ex^4 + dx^3 + cx^2 + bx + a = 0 \text{ has roots } \frac{1}{\alpha}, \frac{1}{\beta}, \frac{1}{\gamma}, \frac{1}{\delta}.$$

One can verify this phenomenon with an example such as $x^2 - 3x - 10 = 0$, whose roots are 5 and −2. We then compare it to the solution of the quadratic equation $10x^2 - 3x - 1 = 0$, whose roots are $\frac{1}{5}$ and $-\frac{1}{2}$. This astonishing relationship should have been part of the elementary algebra course, but it was never mentioned!

An Oddity: The Russian Peasant's Method of Multiplication

It is said that Russian peasants used a rather odd, perhaps even primitive, method to multiply two numbers. As cumbersome as this algorithm is, it requires very little mathematics. Consider this example where we are to find the product of 43×92. We begin by setting up a chart of two columns with the two factors in the first row, as shown in Figure 2-7. The factors of the product we seek are 43 and 92 and head up the columns. One column will be formed by doubling each number to get the next, while the other column will take half the number and drop the remainder. For convenience, the first column will be the doubling column, and the second column will be the halving column. Notice that when we divide in half the odd number 23 (the third number in the second column), we get 11 with a remainder of 1, and we simply drop the 1. The rest of this halving process should now be clear.

43	92
86	46
172	**23**
344	**11**
688	**5**
1376	2
2752	**1**

Figure 2-7

Locate the odd numbers in the halving column (here, the right column), then get the sum of the partner numbers in the doubling column (in this case, the left column). These are highlighted in bold type. This sum gives you the originally required product of 43 and 92. With the Russian peasant's method, we get

$$43 \times 92 = 172 + 344 + 688 + 2752 = 3956.$$

In the example above, we chose to have the first column be the doubling column and the second column be the halving column. We could also have done this by halving the numbers in the first column and doubling those in the second, as can be seen in Figure 2-8.

43	**92**
21	**184**
10	368
5	**736**
2	1472
1	**2944**

Figure 2-8

To complete the multiplication, we find the odd numbers in the halving column (in bold type) and then get the sum of their partner numbers in the doubling column. This gives us

$$43 \times 92 = 92 + 184 + 736 + 2944 = 3956.$$

Although the Russian peasant's method does produce a correct answer, it is clearly not practical in this technological era. However, it should be interesting to observe how this primitive system of arithmetic actually works. Explorations of this kind are not only instructive but entertaining as well.

Here you see what was done in the above multiplication algorithm.

$$
\begin{aligned}
{}^*43\cdot 92 &= (21\cdot 2+1)(92) &&= 21\cdot 184 + 92 = 3956\\
{}^*21\cdot 184 &= (102+1)(184) &&= 10\cdot 368 + 184 = 3864\\
10\cdot 368 &= (5\cdot 2+0)(368) &&= 5\cdot 736 + 0 = 3680\\
{}^*5\cdot 736 &= (2\cdot 2+1)(736) &&= 2\cdot 1472 + 736 = 3680\\
2\cdot 1472 &= (1\cdot 2+0)(1472) &&= 1\cdot 2944 + 0 = 2944\\
{}^*1\cdot 2944 &= (0\cdot 2+1)(2944) &&= 0 + \underline{2944} = 2944\\
& && \qquad\quad 3956
\end{aligned}
$$

For those familiar with the binary system (base 2), one can also explain this Russian peasant method with the following representation:

$$
\begin{aligned}
(43)(92) &= (1\cdot 2^5 + 0\cdot 2^4 + 1\cdot 2^3 + 0\cdot 2^2 + 1\cdot 2^1 + 1\cdot 2^0)(92)\\
&= 2^0\cdot 92 + 2^1\cdot 92 + 2^3\cdot 92 + 2^5\cdot 92\\
&= 92 + 184 + 736 + 2944\\
&= 3956.
\end{aligned}
$$

Whether or not you fully understand the Russian peasant's method of multiplication, you should at least now have a deeper appreciation for the multiplication algorithm you learned in school, even though most people today multiply with a calculator. There are many other multiplication algorithms, yet the one shown here is perhaps one of the strangest. It is through this strangeness that we can appreciate the powerful consistency of mathematics which allows us to conjure up such an algorithm.

Mentally Adding Large Numbers

This method requires taking large numbers and finding a close number which is a multiple of 10. For example, suppose you want to add 786 + 258. The first step is to round them ups to the next multiples of 10, which in this case are 790 and 260, whose sum is 1050. We note the numbers we added to the addends to do the rounding up, and then add them, so we get 4 + 2 = 6. We simply subtract the 6 from 1050 to get 1044, which is the answer to the original addition question: 786 + 258 = 1044.

A Clever Method of Subtracting from 1000

This is a simple technique. It requires that every digit in the minuend (the number you are subtracting from 1000) be subtracted from 9, except for the units digit, which get subtracted from 10. We show this with an example: 1000 – 528.

- Subtract 5 from 9 to get 4.
- Subtract 2 from 9 to get 7.
- Subtract 8 from 10 to get 2.

The answer is 472.

Simplifying Multiplication

When multiplying two numbers where one is even, there is a clever way to simplify the process. Divide the even number in half and double the other number. Suppose you would like to multiply 40 × 233. We begin by taking half the even factor to get 20, and then double the second factor to get 466. We simply multiply 20 × 466 = 9320, which is much easier than the original multiplication. If you want to simplify this further, take half the even number 20, which becomes 10, and multiply it by double the second number 466, which becomes 932, thus making the multiplication trivial, as 10 × 932 = 9320.

An Unusual Method to Multiply Two-Digit Numbers

Let's delve right into this clever and unusual method of multiplying two-digit numbers with some examples and then see why this technique works. Take, for example, the multiplication 95×97. The following steps can be done mentally (with some practice, naturally!):

Step 1:	$95 + 97$	$= 192$
Step 2:	Delete the hundreds digit	$= 92$
Step 3:	Tag two zeros onto the number	$= 9200$
Step 4:	$(100 - 95) \cdot (100 - 97) = 5{\times}3$	$= 15$
Step 5:	Add the last two numbers	$=$ 9215, which is the product being sought!

Here is another example of using this technique to multiply two two-digit numbers:

$$93 \times 96 = \ldots$$

$$93 + 96 = 189$$

~~1~~89 (Delete the hundreds digit)

Tag on two zeros $= 8900$

Then add $(100 - 93) \times (100 - 96) = 7 \times 4 = 28$ to the previously obtained number to get 8928, which is what we sought: 93×96. This technique also works when seeking the product of two numbers that are further apart, as shown in the following example:

$$89 \times 73$$

$$89 + 73 = 162$$

~~1~~62 (Delete the hundreds digit)

Tag on two zeros $= 6200$

Then add $(100 - 89) \times (100 - 73) = 11 \times 27 = 297$ to get 6497, which again leads us to the product of 89×73.

For those who might be curious as to why this unusual technique works, we can use simple algebra that will justify it. We begin with the two two-digit numbers

$$(100 - a) \text{ and } (100 - b) \text{ (where } 0 < a, \text{ and } b < 100)$$

Step 1: $(100 - a) + (100 - b) = 200 - a - b$

Step 2: Delete the hundreds digit, which means subtracting 100 from the number:

$$(200 - a - b) - 100 = 100 - a - b$$

Step 3: Tag on two zeros, which means multiply by 100:

$$(100 - a - b) \times 100 = 10{,}000 - 100a - 100b$$

Step 4: $a \times b$

Step 5: Add the results from steps 3 and 4, and then do some factoring:

$$\begin{aligned} &10{,}000 - 100a - 100b + ab \\ &= 100(100 - a) - (100b - ab) \\ &= 100(100 - a) - b(100 - a) \\ &= (100 - a)(100 - b), \text{ which is what we set out to show.} \end{aligned}$$

Now you just need to practice this technique to master it!

Squaring Multiples of 5

There is a very clever technique for squaring multiples of 5. It simply requires disregarding the 5 in the units digit, and then multiplying the remaining number by that number increased by 1. Tag on 25 to that product to get the answer you seek. Consider as an example the task of finding 65^2. We multiply $6 \times (6+1) = 42$ and then tag on the 25 to get 4225. This can be done with large numbers as well, such as 785^2.

Begin by multiplying $78 \times (78+1) = 6162$ and then just tag on the 25 to get 616,225.

The Old English Method of Multiplication

Over 400 years ago, the English used a very peculiar method to multiply two numbers. They formed the grid shown in Figure 2-9. This procedure is best explained through an example, such 527×839. We begin by placing the first factor along the top of the grid, with each digit over a separate column, and the second number along the right side, with each digit at a row. The next step in the multiplication process is to multiply the number above the first column by the numbers along the right side and placing the two-digit numbers in each rectangle accordingly, that is, $7 \times 8 = 56$, $7 \times 3 = 21$, and $7 \times 9 = 63$.

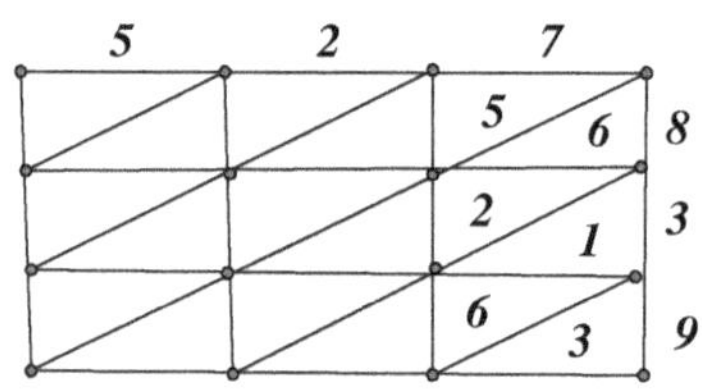

Figure 2-9

The second column on the grid is then filled with the number above the column times each of the numbers to the right of the grid so that it is filled with the numbers 16, 6, and 18, as shown in Figure 2-10.

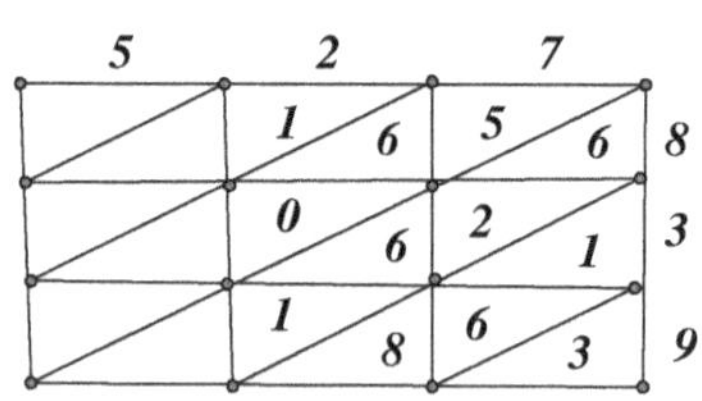

Figure 2-10

The third column is then filled with the product of 5 with each of the numbers to the right of the grid, as shown in Figure 2-11.

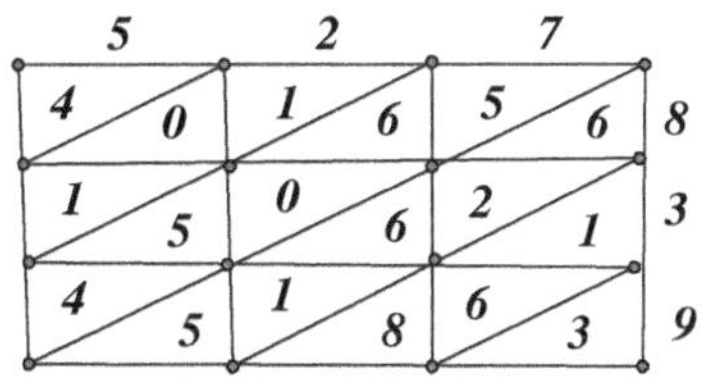

Figure 2-11

Starting with the units digit on the bottom right-hand corner, we add the numbers in the diagonal direction, as shown in Figure 2-12. The first number is simple since it is alone; we place the 3 at the bottom of the first column. We then add the numbers in the next diagonal: 1 + 6 + 8 = 15, so we write the 5 at the bottom of the middle column and carry the 1 to the next diagonal sum, which is 6 + 2 + 6 + 1 + 5 + 1 = 21. We write the 1 at the bottom of the left column and carry the 2 to the next diagonal sum, yielding 5 + 6 + 0 + 5 + 4 + 2 = 22. We place the 2 at the left-side of the bottom row, and once again carry the other 2 to the next diagonal sum: 1 + 0 + 1 + 2 = 4, which goes to the left of the middle row. The last diagonal has one number and that is the 4, which goes to the left of the top row. Thus, we then have the product of 527 × 839 = 442,153, by reading the left-side digits and the bottom list of digits.

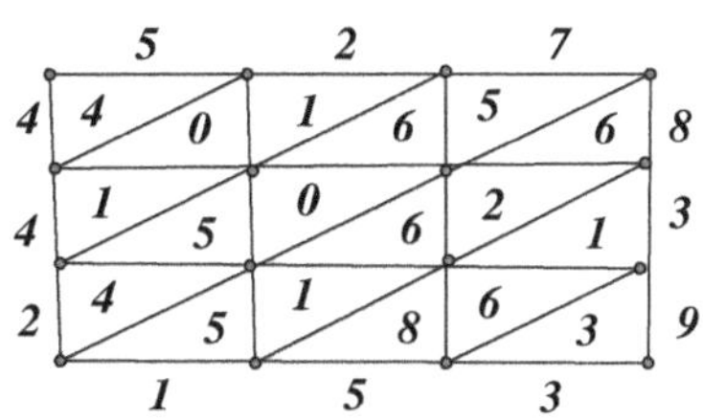

Figure 2-12

Japanese Multiplication

This is a very peculiar form of multiplication; however, it seems to work. The process is that one number is represented by horizontal lines and the other is represented by vertical lines, as you can see in Figure 2-13. The example that we will consider is 23×54. The number 23 has two horizontal lines at the top and three horizontal lines at the bottom. The second number, 54, has five vertical lines on the left and four vertical lines on the right. We then count the points of intersection noted in the figure. To get the product, we take the sum of the top right number and the bottom left number, in this case $8 + 15 = 23$, and sandwich this sum between the top left number (10) and the bottom right number (12). This results in **10** 23 **12**. We then take the two outside digits and add those numbers between them to get 1 (0+23+1) 2, which becomes $1242 = 23 \times 54$.

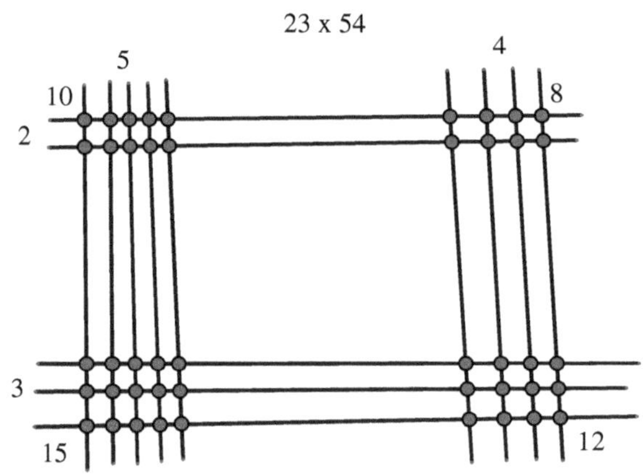

Figure 2-13

Alphametics

One of the great strides made by western civilization was adopting from the civilizations of India and the Islamic world the use of a place-value system for our arithmetic. Working with Roman numerals was not only cumbersome but made many algorithms impossible. The first appearance of the Hindu/Arabic numerals on the European continent was in 1202 in Fibonacci's *Liber Abaci*. Beyond its usefulness, the place-value system can also provide us with some recreational mathematics that can stretch our understanding of and facility with the place-value system and the algorithms it enables.

Applying reasoning skills to an addition algorithm situation can be entertaining. Consider the following, where the letters represent the digits of a simple addition:

$$\begin{array}{rccccc} & & \mathbf{S} & \mathbf{E} & \mathbf{N} & \mathbf{D} \\ . & & \mathbf{M} & \mathbf{O} & \mathbf{R} & \mathbf{E} \\ \hline & \mathbf{M} & \mathbf{O} & \mathbf{N} & \mathbf{E} & \mathbf{Y} \end{array}$$

Find the digits that are represented by the letters to make this addition correct, and show that the solution is unique. Most important in this activity is the analysis, and particular attention should be given to the reasoning used. At first, this may appear to be a daunting task, but if you take it step-by-step, you will find it rewarding and entertaining.

The sum of two four-digit numbers cannot yield a number greater than 19,999. Therefore, **M** = **1**.

We then have MORE < 2000 and SEND < 10,000. It follows that MONEY < 12,000. Thus, O can be either 0 or 1. But the 1 is already used; therefore, **O** = **0**.

We now have:

$$\begin{array}{rccccc} & & \mathbf{S} & \mathbf{E} & \mathbf{N} & \mathbf{D} \\ . & & \mathbf{1} & \mathbf{0} & \mathbf{R} & \mathbf{E} \\ \hline & \mathbf{1} & \mathbf{0} & \mathbf{N} & \mathbf{E} & \mathbf{Y} \end{array}$$

Now, MORE < 1100. If SEND were less than 9000, then MONEY < 10,100, which would imply that N = 0. But this cannot be, since 0 was already used; therefore, SEND > 9000, so that **S** = **9**. We now have:

$$\begin{array}{rccccc} & & \mathbf{9} & \mathbf{E} & \mathbf{N} & \mathbf{D} \\ . & & \mathbf{1} & \mathbf{0} & \mathbf{R} & \mathbf{E} \\ \hline & \mathbf{1} & \mathbf{0} & \mathbf{N} & \mathbf{E} & \mathbf{Y} \end{array}$$

The remaining digits with which we may complete the problem are $\{2,3,4,5,6,7,8\}$.

Let us examine the units digits. The greatest sum is $7 + 8 = 15$, and the least sum is $2 + 3 = 5$.

If $D + E < 10$, then $D + E = Y$ with no carry over into the tens column. Otherwise, $D + E = Y + 10$, with a 1 carried over to the tens column.

Taking this argument one step further to the tens column, we get $N + R = E$, with no carryover, or $N + R = E + 10$, with a carryover of 1 to the hundreds column. However, if there is no carryover to the hundreds column, then $E + 0 = N$, which implies that $E = N$. This is not permissible. Therefore, there must be a carryover to the hundreds column, so, $N + R = E + 10$, and $E + 0 + 1 = N$, or $E + 1 = N$. Substituting this value for N into the previous equation we get: $(E + 1) + R = E + 10$, which implies that $R = 9$. But this has already been used for the value of S. We must try a different approach. We shall assume, therefore, that $D + E = Y + 10$, since we apparently need a carryover into the tens column, where we just reached a dead end.

Now the sum in the tens column is $1 + 2 + 3 < 1 + N + R < 1 + 7 + 8$. If, however, $1 + N + R < 10$, there will be no carryover to the hundreds column, leaving the previous dilemma of $E = N$, which is not allowed. We then have $1 + N + R = E + 10$, which ensures the necessary carryover to the hundreds column. There, $1 + E + 0 = N$, or $E + 1 = N$.

Substituting this in the above equation $(1 + N + R = E + 10)$ gives us $1 + (E + 1) + R = E + 10$, or $\mathbf{R = 8}$. We now have:

$$\begin{array}{ccccc} & 9 & E & N & D \\ . & 1 & 0 & 8 & E \\ \hline 1 & 0 & N & E & Y \end{array}$$

From the remaining list of available digits, we find that $D + E < 14$. So from the equation $D + E = Y + 10$, Y is either 2 or 3. If $Y = 3$, then $D + E = 13$, implying that the digits D and E can take on only 6 or 7. If $D = 6$ and $E = 7$, then from the previous equation $E + 1 = N$, we would have $N = 8$, which is unacceptable since $R = 8$. If $D = 7$ and $E = 8$, then from the previous equation $E + 1 = N$, we would have $N = 9$, which is unacceptable since $S = 9$. Therefore, $\mathbf{Y = 2}$. We now have:

$$\begin{array}{r} 9\ E\ N\ D \\ .\quad 1\ 0\ 8\ E \\ \hline 1\ 0\ N\ E\ 2 \end{array}$$

Thus, D + E = 12. The only way to get this sum is with 5 and 7. If E = 7, we once again get from E + 1 = N the contradictory N = 8, which is not acceptable. Therefore, **D** = **7** and **E** = **5**. We can now again use the equation E + 1 = N to get **N** = **6**. Finally, we get the solution:

$$\begin{array}{r} 9\ 5\ 6\ 7 \\ .\quad 1\ 0\ 8\ 5 \\ \hline 1\ 0\ 6\ 5\ 2 \end{array}$$

Suppose your son at college is asking for more funds and he sends the cryptic message above, knowing that you like puzzles. However, you decide you'd like to send him a message in the same style, saying:

$$\begin{array}{r} U\ S\ E \\ L\ E\ S\ S \\ \hline S\ O\ N\ N\ Y \end{array}$$

However, you would like to put this message in the same numerical style and have him figure out what the message says. You have a choice of three possibilities, which we will generate now.

As we begin, we notice that the S should be a 1, the L a 9, and the O should be a 0. Furthermore, since we have two Ss added to get an N, and 1 + 1 = 2, the N should be a 2. The remaining letters can be done by trial and error. For example, E could not be 3 because then U would have to be 9, which was already used. Therefore, we only have 4, 5, or 7 available to use as E.

$$\begin{array}{r} U\ 1\ E \\ 9\ E\ 1\ 1 \\ \hline 1\ 0\ 2\ 2\ Y \end{array}$$

As we mentioned earlier, there are three possible solutions to this problem. In other words, the father can send any one of these three to get the message to his son, and they are

$$\begin{array}{r} 814 \\ 9411 \\ \hline 10225 \end{array} \qquad \begin{array}{r} 715 \\ 9511 \\ \hline 10226 \end{array} \qquad \begin{array}{r} 517 \\ 9711 \\ \hline 10228 \end{array}$$

Now the father has to hope that the student gets the message.

Chapter 3

Logical Reasoning

An Impressive Clock Oddity

This impressive clock oddity is quite entertaining and open to some speculation as to why it works. We begin by considering the normal clock, but instead of using the normal numerals from 1 to 12, we will place a 3 at the usual 1 position, and then in sequence place multiples of 3 at each successive position (clockwise) until we get to the 12 position. Then besides these multiples of 3, we repeat this procedure by placing a 4 at the 1 position and continue placing multiples of 4 until we get to the 12 position again. We show this in Figure 3-1.

We are now ready to be entertained by this oddity and arrangement of numbers. First, by subtracting each pair of numbers, you will have gotten the hour position at each place. For example, the 5-hour position can be obtained by subtracting $20 - 15 = 5$. If you take the difference of the pairs of numbers and add that to the larger of the two numbers, you will get the number of minutes to that position. For example, the 25-minute position can be obtained by subtracting $20 - 15 = 5$ and adding 5 to the larger of the pair, namely, $20 + 5 = 25$. As an extra bonus, if you take the square root of the sum of the squares of each pair of numbers, you will get the number of minutes represented at that position. We can see that by taking $20^2 + 15^2 = 625 = 25^2$. You might like to try these functions at some other positions on the face

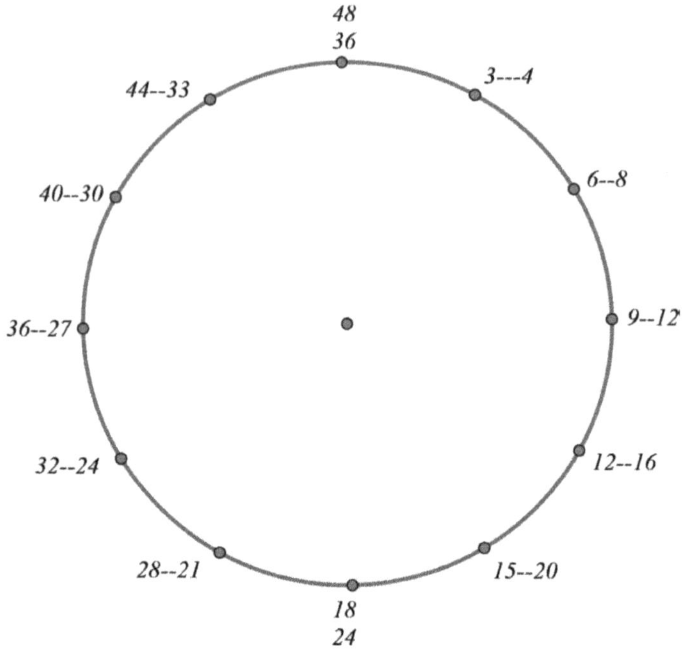

Figure 3-1

of the clock to become convinced of this marvelous relationship. This oddity will certainly be an impressive form of entertainment and provoke thoughts about why the numbers behave the way they do.

The Angles of the Hands of a Clock

We take for granted that the angles made by the hands of a clock are not consistent and change minute-by-minute, hour-by-hour. And yet, the clock can be an interesting source of mathematical applications. These applications are focused on mathematics, rather than the more common application of mathematics to other disciplines. That is, the clock can be a good source for some entertainment with mathematics.

For starters, we might want to consider the *exact time* that the hands of a clock will overlap after 4:00. An off-the-cuff response might be 4:20. Unfortunately, the hands of the clock move continuously, and the minute hand moves faster than the hour hand. With this in mind, the astute reader might estimate the answer to be between 4:21 and 4:22. Keep in mind that the hour hand moves through the interval between minute markers every 12 minutes. Therefore, it will leave the interval 4:21–4:22 at 4:24. This, however, doesn't answer the original question about the exact time of this overlap.

We will provide a "trick" to help with this situation. Realizing that this first guess of 4:20 is not the correct answer, since the hour hand does not remain stationary and moves when the minute hand moves, the trick is to simply multiply 20 (the wrong answer) by $\frac{12}{11}$ to get $21\frac{9}{11}$, which yields the correct answer: $4{:}21\frac{9}{11}$.

One way to understand the movement of the hands of a clock is that the hands travel independently around the clock at uniform speeds. The minute markings on the clock (henceforth referred to as "markers") serve to denote distance as well as time. An analogy could be drawn here to the "uniform motion" of automobiles, where a fast car overtakes a slower one, a popular visualization in elementary algebra courses. Here, the analogous situation would be to find the distance necessary for a car traveling at 60 m.p.h. to overtake a car with a head start of 20 miles and traveling at 5 m.p.h.

Now consider 4 o'clock as the initial time on the clock. Our problem will be to determine exactly when the minute hand will overtake the hour hand after 4 o'clock. Consider the speed of the hour hand to be r, then the speed of the minute hand must be $12r$. We seek the distance, measured by the number of markers traveled, that the minute hand must travel to overtake the hour hand. Let us refer to this distance as d markers. Hence, the distance that the hour hand travels is $d - 20$ markers, since it has a 20-marker head start over the minute hand. (See Figure 3-2.)

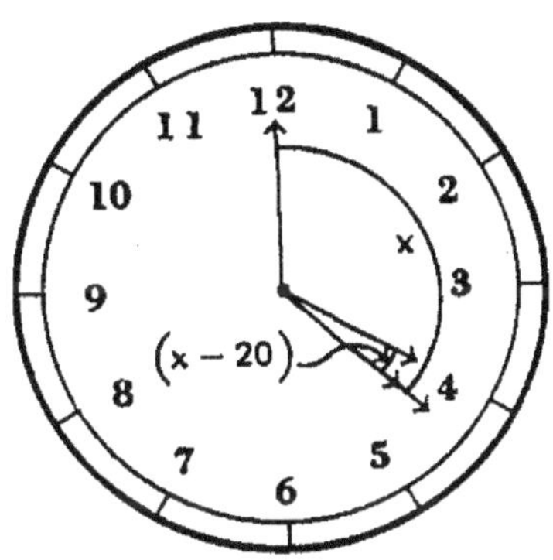

Figure 3-2

For this to take place, the times required for the minute hand, $\frac{d}{12r}$, and for the hour hand, $\frac{d-20}{r}$, are the same. Therefore, $\frac{d}{12r} = \frac{d-20}{r}$, and $d = \frac{12}{11} \cdot 20 = 21\frac{9}{11}$. Thus, the minute hand will overtake the hour hand at exactly $4{:}21\frac{9}{11}$.

Consider the expression $d = \frac{12}{11} \cdot 20$. The quantity 20 is the number of markers that the minute hand had to travel to get to the desired position, if we assume the hour hand remained stationary. Obviously, the hour hand does not remain stationary. Hence, we must multiply this quantity by $\frac{12}{11}$, since the minute hand must travel $\frac{12}{11}$ as far. Let us refer to this fraction, $\frac{12}{11}$, as the *correction factor*.

To familiarize yourself with the correction factor, choose some short and simple problems. For example, you may want to find the exact time when the hands of a clock overlap between 7 and 8 o'clock. Here you would first determine how far the minute hand would have to travel from the "12" position to the position of the hour hand, assuming again that the hour hand remains stationary. Then by multiplying the number of markers, 35, by the correction factor, $\frac{12}{11}$, you will obtain the exact time, $7{:}38\frac{2}{11}$, that the hands will overlap.

To enhance your understanding of this new procedure, consider a person checking a wristwatch against an electric clock and noticing that the hands on the wristwatch overlap every 65 minutes (as measured by the electric clock). Is the wristwatch fast, slow, or accurate?

You may wish to consider the problem in the following way. At 12 o'clock, the hands overlap exactly. Using the previously described method, we find that the hands will again overlap at exactly $1{:}05\frac{5}{11}$, and then again at exactly $2{:}10\frac{10}{11}$, and again at exactly $3{:}16\frac{4}{11}$ and so on. Each time there is an interval of $65\frac{5}{11}$ minutes between overlapping positions. Hence, the person's watch is inaccurate by $\frac{5}{11}$ of a minute. Can you now determine if the wristwatch is fast or slow?

There are many other interesting, and sometimes rather difficult, problems made simple by this "correction factor." You may easily develop your own problems. For example, you may wish to find the exact times when the hands of a clock will be perpendicular (or form a straight angle) between, say, 8 and 9 o'clock.

Again, you would try to determine the number of markers that the minute hand would have to travel from the "12" position until it forms the desired angle with the stationary hour hand. Then multiply this number by the correction factor, $\frac{12}{11}$, to obtain the exact actual time. That is, to find the exact time that the hands of a clock are *first* perpendicular between 8 and 9 o'clock, determine the desired position of the minute hand when the hour hand remains stationary. This would be on the 25-minute marker. Then, multiply 25 by $\frac{12}{11}$ to get $8{:}27\frac{3}{11}$, the exact time when the hands are *first* perpendicular after 8 o'clock.

For those who want to look at this issue from a non-algebraic viewpoint, you could justify the $\frac{12}{11}$ correction factor for the interval between overlaps in the following way. Think of the hands of a clock at noon. During the next 12 hours (i.e., until the hands reach the same position at midnight) the hour hand makes one revolution, the minute hand makes 12 revolutions, and the minute hand coincides with the hour hand 11 times (including midnight, but not noon, starting just after the hands separate at noon). Since each hand rotates at a uniform rate, the hands overlap each $\frac{12}{11}$ of an hour, or $65\frac{5}{11}$ minutes. This can be extended to other situations.

You should derive a great sense of achievement and enjoyment from employing this simple procedure to solve what usually appears to be a very difficult clock problem.

Birthday Matches

Mathematics tends to provide counterintuitive situations. Some of these will leave the uninitiated in awe. One of the best ways to convince others of the "power" of probability is to exhibit such examples. The results of this oddity, aside from being entertaining, will upset your sense of intuition.

Let us suppose you are in a group of about 35 people. What do you think the chances (or probability) are of two members of the group having the same birth date (month and day, only). Intuitively, one usually begins to think about the likelihood of two people having the same date out of a selection of 365 days (assuming no leap year). Perhaps 2 out of 365? That would be a probability of $\frac{2}{365} = .005479 \approx \frac{1}{2}\%$. A minuscule chance.

Let's consider the "randomly" selected group of the first 35 presidents of the United States. You may be astonished that there are two with the same birth date:

the 11th president, James K. Polk (November 2, 1795), and

the 29th president, Warren G. Harding (November 2, 1865).

You may be surprised to learn that for a group of 35, the probability that two members will have the same birth date is greater than 8 out of 10, or $\frac{8}{10} = 80\%$.

If you have the opportunity, you may wish to try your own experiment by selecting 10 groups of about 35 members each to check on birthday matches. For groups of 30, the probability that there will be a match is greater than 7 out of 10; that is, in 7 of these 10 rooms there ought to be a match of birth dates. What causes this incredible and unanticipated result? Can this be true? It is contrary to our intuition. To relieve you of your curiosity, we will consider the situation in detail.

Let's consider a class of 35 students. What is the probability that one selected student matches his own birth date? Clearly, it is a *certainty*, that is, the chances are 1 in 1. This can be written as $\frac{365}{365}$.

The probability that another student does *not* match the first student is

$$\frac{365-1}{365} = \frac{364}{365}.$$

The probability that a third student does *not* match the first and second students is

$$\frac{365-2}{365} = \frac{363}{365}.$$

The probability of all 35 students *not* having the same birth date is the product of these probabilities:

$$p = \frac{365}{365} \cdot \frac{365-1}{365} \cdot \frac{365-2}{365} \cdot \ldots \cdot \frac{365-34}{365}.$$

Since the probability that two students in the group have the same birth date (q) and the probability that two students in the group do not have the same birth date (p) is a certainty, the sum of those probabilities must be 1. Thus, $p + q = 1$. In this case,

$$q = 1 - \frac{365}{365} \cdot \frac{365-1}{365} \cdot \frac{365-2}{365} \cdot \ldots \cdot \frac{365-33}{365} \cdot \frac{365-34}{365} \approx .8143832388747152.$$

In other words, the probability that there will be a birth date match in a randomly selected group of 35 people is somewhat greater than $\frac{8}{10}$. This is quite unexpected when one considers there were 365 dates from which to choose.

In Figure 3-3, we show the probability of birthday matches in groups of various sizes. Counterintuitively, we notice that a group of 55 people will have an almost 100% chance of having a birthday match. When one considers that there are 365 possible dates, this result is unbelievable... But true! Notice how quickly almost-certainty is reached. With about 60 students in a room, the chart indicates that it is almost certain (.99) that two students will have the same birth date.

Number of People in Group	Probability of a Birth Date Match
10	.1169481777110776
15	.2529013197636863
20	.4114383835805799
25	.5686997039694639
30	.7063162427192686
35	.8143832388747152
40	.891231809817949
45	.9409758994657749
50	.9703735795779884
55	.9862622888164461
60	.994122660865348
65	.9976831073124921
70	.9991595759651571

Figure 3-3

Were one to do this with the death dates of the first 35 presidents, one would notice that two died on March 8th (Millard Fillmore in 1874 and William H. Taft in 1930) and three presidents died on July 4th (John Adams and Thomas Jefferson in 1826, and James Monroe in 1831). Above all, this astonishing demonstration should reveal the inadvisability of relying too much on intuition.

The Odd Flight of the Bumblebee

There are challenges that seem unrealistic and impossible to solve. However, we can use clever techniques to solve some problems. The problem presented here might remind you of a topic in elementary algebra, and you might be tempted to use the procedures you learned there. That would be a mistake. When you see how to solve this problem, it will be an eye-opener and a good example for future such encounters. Don't be deterred by the length of the problem. You will be delighted (and entertained) with the unexpected simplicity of the solution.

> *Two trains, serving the Chicago to New York route, a distance of 800 miles, start towards each other at the same time, along the same tracks. One train is traveling uniformly at 60 miles per hour and the other at 40 miles per hour. At the same time, a bumblebee begins to fly from the front of one of the trains at a speed of 80 miles per hour towards the oncoming train. After touching the front of this second train, the bumblebee reverses direction and flies towards the first train (still at the same speed of 80 miles per hour). The bumblebee continues this back and forth flying until the two trains collide, crushing the bumblebee. How many miles did the bumblebee fly before its demise?*

It is natural to try and find the total distance that the bumblebee traveled. An immediate reaction is to set up an equation based on the famous relationship "rate times time equals distance." However, this back-and-forth path is difficult to determine and requires considerable calculation. Just the notion of having to do this can cause serious frustration. Even if you were able to determine each of the parts of the bumblebee's flight, it is still very difficult to solve the problem in this way.

A much more elegant approach would be to solve a simpler analogous problem. One might also consider looking at the problem from a different point of view. We seek to find the *distance* the bumblebee traveled. If we knew the *time* the bumblebee traveled, we could determine the bumblebee's distance because we already know the *speed* of the bumblebee. Having two parts of the equation "rate × time = distance" will provide the third part. So having the *time* and the *speed* will yield the distance traveled, albeit in various directions.

The time the bumblebee traveled can be easily calculated, since it traveled the entire time the two trains were traveling towards each other until they collided. To determine the time, t, that the trains traveled, we set up an equation as follows. The distance of the first train is $60t$ and the distance of the second train is $40t$. The total distance the two trains traveled is 800 miles. Therefore, $60t + 40t = 800$, so $t =$ 8 hours, which is also the time the bumblebee traveled. We can now find the distance the bumblebee traveled using the relationship rate × time = distance, which gives us $(8)(80) = 640$ miles.

It is important to avoid falling into the trap of trying to do what the problem calls for directly. At times, a more circuitous method is much more efficient. Lots can be learned from this solution. You see, dramatic solutions are often more useful than traditional solutions, since they provide an opportunity "to think out of the box."

A Logical Arrangement

Suppose you have 12 Xs and you're challenged to place them into a square matrix of 36 boxes, shown in Figure 3-4, such that you have six lines, horizontal, vertical or diagonal with 4 Xs in each line. X, X, X, X, X, X, X, X, X, X, X, X. Before looking at Figure 3-5, see if you can solve the problem posed here.

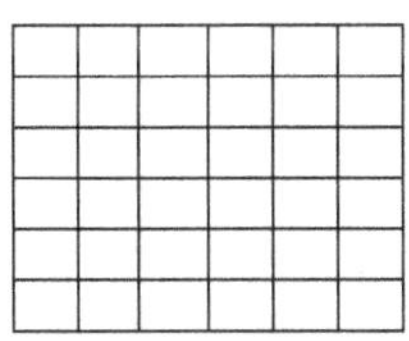

Figure 3-4

Although it may be a bit puzzling, solving the challenge is also rewarding. One possible solutions is shown in Figure 3-5.

			X		
	X	X	X	X	
	X		X		
X	X	X	X		
	X				

Figure 3-5

Unusual Aspects of a Hexagram

The challenge here is to place the numbers 1–12 at the points of intersection in the hexagram shown in Figure 3-6, so that every straight line has a sum of 26 and the sum of the numbers on the vertices of the hexagram is also 26. Once again, try to meet the challenge without looking at Figure 3-7 below.

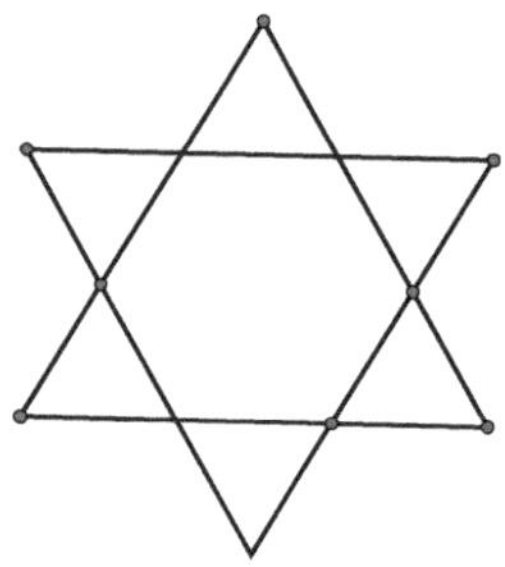

1, 2, 3, 4, 5, 6, 7, 8, 9, 10, 11, 12

Figure 3-6

After various trials and errors, Figure 3-7 shows three solutions to the challenge presented.

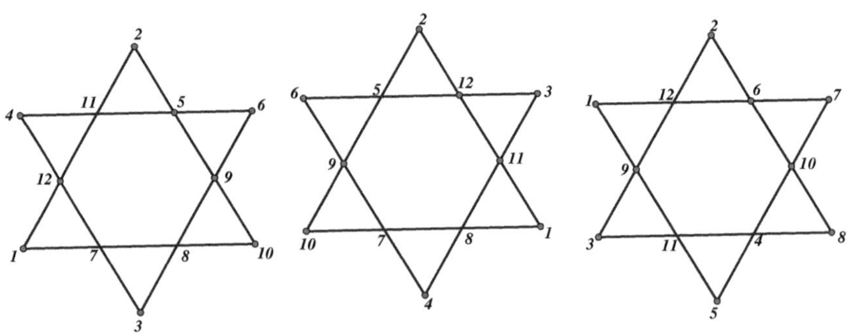

Figure 3-7

Calendar Marvels

There are unexpected relationships among the numbers of any calendar month. Consider the month of May shown in Figure 3-8. When a rectangular arrangement (not a square) is indicated, we find that the sum of the dates at the opposite vertices are equal. In the smaller of these two rectangles, we find that 6 + 15 = 21 = 8 + 13. In the larger rectangle, we have 11 + 30 = 41 = 25 + 16. This can be done with any size of rectangle, as it is a peculiarity of the calendar arrangement.

Figure 3-8

In a 3 × 3 square, shown in Figure 3-9, we find that not only are the sums of the opposite vertices equal (as is the case here: 8 + 24 = 32 = 10 + 22), but with this special size of square, the sum of the center row and center column are also equal: 15 + 16 + 17 = 48 = 9 + 16 + 23. Also, the sum of the numbers in each diagonal is the same: 8 + 16 + 24 = 48 = 10 + 16 + 22.

When we have other square arrangements, the sum of numbers along each diagonal is the same. In the 2 × 2 square shown in Figure 3-9, we have 5 + 13 = 18 = 6 + 12. It becomes more dramatic when we consider the 4×4 square shown in Figure 3-9, where the sum of the diagonals is once again the same: 4 + 12 + 20 + 28 = 64 = 7 + 13 + 19 + 25. A motivated reader might want to show that rectangular and square arrangements of numbers follow these patterns regardless of which month is selected.

Figure 3-9

Beautiful Magic Squares

Magic squares are square arrangement of numbers where the sums of the numbers in each row, each column, and each diagonal are the same. Magic squares have fascinated folks for centuries.

The typical way to construct a 4×4 magic square is to take the numbers from 1 to 16 and place them as shown in Figure 3-10, then replace the numbers in the diagonals with their complement. The complement is the number which when added to it equals 17.

1	2	3	**4**
5	**6**	**7**	8
9	**10**	**11**	12
13	14	15	**16**

Figure 3-10

This produces the magic square seen in Figure 3-11. Clearly, as all the rows, columns and diagonals have a sum of 34, this truly has produced a magic square.

16	2	3	**13**
5	**11**	**10**	8
9	**7**	**6**	12
4	14	15	**1**

Figure 3-11

There is one magic square that stands out from among the rest for its beauty. Its many unusual properties go beyond those required for a square matrix of numbers to be considered "magic." This magic square even comes to us through art, and not through the usual mathematical channels. It is depicted in the background of a famous engraving produced in 1514 by renowned German artist Albrecht Dürer (1471–1528), who lived in Nürnberg, Germany. (See Figure 3-12.)

Figure 3-12

Dürer signed most of his works with his initials, A over D, with the year in which the work was made. Here, we find it near the lower right corner of the picture. We notice that it was made in the year 1514, as we can see from the magnification in Figure 3-13.

Figure 3-13

The observant reader may notice that the two center cells of the magic square's bottom row depict the year as well. The incredible thing is that when we construct a magic square using the standard algorithm for doing so, the two middle columns are interchanged. In other words, Dürer changed the two middle columns to get the year to appear in the two center cells of the last row. Let us look at this magic square more closely in Figure 3-14.

First, let's make sure that it is a magic square. The sum of all the rows and all the columns must be equal. In fact, they all add to 34. That is all that would be required for this square matrix of numbers to be considered a "magic square." However, Dürer's magic square has lots more properties that other magic squares do not have. We shall list some here:

- The four corner numbers have a sum of 34.

$$16 + 13 + 1 + 4 = 34.$$

Figure 3-14

- Each of the four corner 2×2 squares has a sum of 34.

$$16 + 3 + 5 + 10 = 34$$
$$2 + 13 + 11 + 8 = 34$$
$$9 + 6 + 4 + 15 = 34$$
$$7 + 12 + 14 + 1 = 34.$$

- The center 2×2 square has a sum of 34.

$$10 + 11 + 6 + 7 = 34.$$

- The sum of the numbers in the diagonal cells equals the sum of the numbers in the non-diagonal cells.

$$16 + 10 + 7 + 1 + 4 + 6 + 11 + 13 =$$
$$3 + 2 + 8 + 12 + 14 + 15 + 9 + 5 = 68.$$

- The sum of the squares of the numbers in the diagonal cells equals the sum of the squares of the numbers in the non-diagonal cells.

$$16^2 + 10^2 + 7^2 + 1^2 + 4^2 + 6^2 + 11^2 + 13^2 =$$
$$3^2 + 2^2 + 8^2 + 12^2 + 14^2 + 15^2 + 9^2 + 5^2 = 748.$$

- The sum of the cubes of the numbers in the diagonal cells equals the sum of the cubes of the numbers in the non-diagonal cells.

$$16^3+10^3+7^3+1^3+4^3+6^3+11^3+13^3=$$
$$3^3+2^3+8^3+12^3+14^3+15^3+9^3+5^3=9248.$$

- The sum of the squares of the numbers in both diagonal cells equals the sum of the squares of the numbers in the first and third rows.

$$16^2+10^2+7^2+1^2+4^2+6^2+11^2+13^2=$$
$$16^2+3^2+2^2+13^2+9^2+6^2+7^2+12^2=748.$$

- The sum of the squares of the numbers in both diagonal cells equals the sum of the squares of the numbers in the second and fourth rows.

$$16^2+10^2+7^2+1^2+4^2+6^2+11^2+13^2=$$
$$5^2+10^2+11^2+8^2+4^2+15^2+14^2+1^2=748.$$

- The sum of the squares of the numbers in both diagonal cells equals the sum of the squares of the numbers in the first and third columns.

$$16^2+10^2+7^2+1^2+4^2+6^2+11^2+13^2=$$
$$16^2+5^2+9^2+4^2+2^2+11^2+7^2+14^2=748.$$

- The sum of the squares of the numbers in both diagonal cells equals the sum of the squares of the numbers in the second and fourth columns.

$$16^2+10^2+7^2+1^2+4^2+6^2+11^2+13^2=$$
$$3^2+10^2+6^2+15^2+13^2+8^2+12^2+1^2=748.$$

- Notice the following beautiful symmetries:

$$2+8+9+15=3+5+12+14=34$$
$$2^2+8^2+9^2+15^2=3^2+5^2+12^2+14^2=374$$
$$2^3+8^3+9^3+15^3=3^3+5^3+12^3+14^3=4624.$$

- The sum of each adjacent upper and lower pair of numbers (vertically) produces a pleasing symmetry:

16 + 5 = **21**	3 + 10 = **13**	2 + 11 = **13**	13 + 8 = **21**
9 + 4 = **13**	6 + 15 = **21**	7 + 14 = **21**	12 + 1 = **13**

- The sum of each adjacent upper and lower pair of numbers (horizontally) produces a pleasing symmetry:

16 + 3 = **19**	2 + 13 = **15**
5 + 10 = **15**	11 + 8 = **19**
9 + 6 = **15**	7 + 12 = **19**
4 + 15 = **19**	14 + 1 = **15**

Besides the Dürer's magic square, there are 4 × 4 squares that can be generated out of a 7 × 7 square arrangement of numbers, as shown in Figure 3-15.

4	9	6	15	4	9	6
14	7	12	1	14	7	12
11	2	13	8	11	2	13
5	16	3	10	5	16	3
4	9	6	15	4	9	6
14	7	12	1	14	7	12
11	2	13	8	11	2	13

Figure 3-15

When one places a 4 × 4 framework anywhere on this array of numbers, one ends up with a magic square where the rows, columns, and diagonals have the sum of 34. We show in Figure 3-16 one random 4×4 arrangement of numbers which will yield that magic number 34 in each row, column, and diagonal.

4	9	6	15	4	9	6
14	7	12	1	14	7	12
11	2	13	8	11	2	13
5	16	3	10	5	16	3
4	9	6	15	4	9	6
14	7	12	1	14	7	12
11	2	13	8	11	2	13

Figure 3-16

Most Unusual Pythagorean Magic Squares

In Figure 3-17, have two magic squares on the legs of a right triangle and one magic square on the hypotenuse. First of all, each of these magic squares has the same sums in its rows, columns, and diagonals. What makes this astonishing is that the corresponding positions of the two leg magic squares and the hypotenuse magic square follow the Pythagorean theorem. For example, consider the top right corner of each of the three magic squares. We have $18^2 + 24^2 = 30^2$. Another example: if we take the center square of each of these three magic squares, we find that $15^2 + 20^2 = 25^2$.

This can be extended to taking the sum of squares of corresponding rows, columns, or diagonals. For instance, consider the sum of the diagonals squared: $(18 + 15 + 12)^2 + (24 + 20 + 16)^2 = (30 + 25 + 20)^2$. You may want to check other columns or rows to see that this still holds true. Moreover, the sum of the numbers in the hypotenuse squared equals the sum of the numbers in leg A squared plus the sum of the numbers in leg B squared. That is, $(225)^2 = (135)^2 + (180)^2$, or $50{,}625 = 18{,}225 + 32{,}400$.

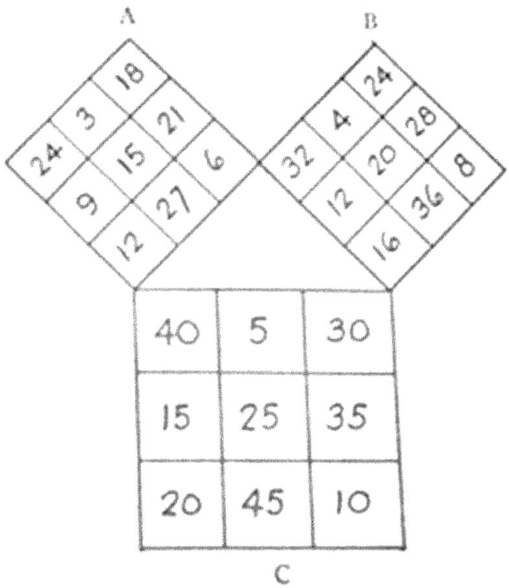

Figure 3-17

A Palindromic Magic Square

The smallest magic square consists of nine cells, as shown in Figure 3-18, where the sum of the numbers in the rows, columns, and diagonals is 15.

4	9	2
3	5	7
8	1	6

Figure 3-18

This particular nine-cell magic square has a unique property. If we consider the three-digit numbers formed by the numbers in the rows in both directions, we get the following amazing result:

$$492^2 + 357^2 + 816^2 = 1{,}035{,}369 = 294^2 + 753^2 + 618^2.$$

Now consider the columns, where the same amazing phenomenon results:

$$438^2 + 951^2 + 276^2 = 1{,}172{,}421 = 834^2 + 159^2 + 672^2.$$

Note how the sums of squares in each case are reversals of one another.

This 3 × 3 magic square can also create an unusual magic square with palindromic numbers. We take the original magic square in Figure 3-19a and flip it to get the magic square seen in Figure 3-19b. Then we take the numbers in the second magic square and sandwich them around the numbers in the corresponding positions of the first magic square. For example, for the top right square we take the 8 from the second magic square and place it on both sides of the 2 in the first magic square to get 828. In the middle square on the top row, we place the 1 on either side of the 9 to get 191. When we do this to all the other cells, we get the magic square shown in Figure 3-19c, where all the elements are palindromic numbers, and the sums of all the rows, columns, and diagonals are equal to 1665. This is an amazing palindromic magic square.

4	9	2
3	5	7
8	1	6

Figure 3-19a

6	1	8
7	5	3
2	9	4

Figure 3-19b

646	191	828
737	555	373
282	919	464

Figure 3-19c

A Puzzling Paradox

A farmer had 17 sheep and wanted to give them to his three sons. The eldest was to get $\frac{1}{2}$ of the sheep, the middle son was to get $\frac{1}{3}$ of the sheep, and the youngest son was to get $\frac{1}{9}$ of the sheep. They could not figure out how many sheep each son was to get because

$$\frac{1}{2}+\frac{1}{3}+\frac{1}{9}\neq 1.$$

They asked a neighboring farmer if he could help solve the problem. He offered to give them one additional sheep so that now they had 18 sheep and could make the allocation. So the eldest son was able to get $\frac{1}{2}$ of the sheep, which was 9, the second oldest son was able to get $\frac{1}{3}$ of the sheep, which was 6, and the youngest son was able to get $\frac{1}{9}$ of the sheep which was 2 sheep. In total, the sons will have gotten $9 + 6 + 2 = 17$ sheep. The kind neighboring farmer takes back his one sheep, as the sons now have the correct allotment. How is it possible that this sheep made the difference and then was not needed? A challenge for the reader.

A Clever Strategy—Anticipating Heads and Tails

We will consider how some clever reasoning along with elementary algebraic knowledge will help you solve a problem that seems impossibly difficult.

Consider the following problem.

You are seated at a table in a dark room. On the table there are 12 pennies, 5 of which are heads up and 7 are tails up. You know where the coins are, so you can move or flip any coin, but because it is dark you will not know if the coin you are touching was originally heads up or tails up. You are to separate the coins into two piles (possibly flipping some of them) so that when the lights are turned on there will be an equal number of heads in each pile.

Your first reaction is "you must be kidding!" "How can anyone do this task without seeing which coins are heads or tails up?" This is where a clever (yet incredibly simple) use of algebra will be the key to the solution.

Here is what you do. (You might actually want to try it with 12 coins.) Separate the coins into two piles of 5 and 7 coins each. Then flip over the coins in the smaller pile. Now both piles will have the same number of heads! That's all! You will think this is magic. How did this happen? Well, this is where algebra helps us understand what was actually done.

Let's say that when you separate the coins in the dark room, h heads will end up in the 7-coin pile. Then the other pile, the 5-coin pile, will have $5 - h$ heads and $5 - (5 - h) =$ tails. When you flip all the coins in the smaller pile, the $5 - h$ heads become tails, and the h tails become heads. Now each pile contains h heads!

A Clever Arrangement Trick

In Figure 3-20, we have six separate sticks, and we need to arrange them so that each stick touches every other stick in the set.

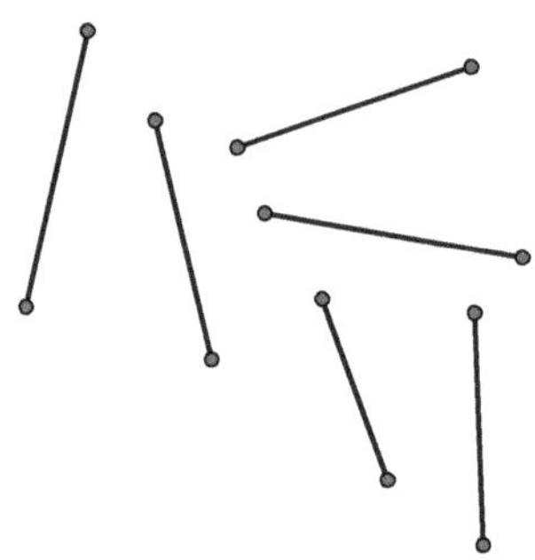

Figure 3-20

The typical response to the question is the arrangement shown in Figure 3-21. However, that would not be correct because the sticks have a certain thickness, and they would not all be touching. Generally, the sticks would touch those adjacent to them but not necessarily all opposite sticks. Before looking further at Figure 3-22, the reader should try to find a solution.

Thus, an arrangement where each stick touches each of the other sticks would be one shown in Figure 3-22.

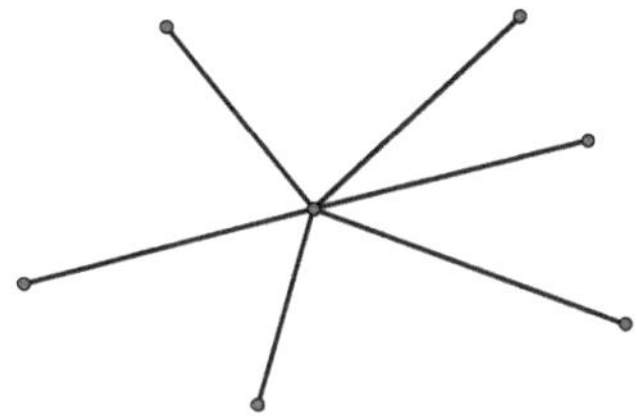

Figure 3-21

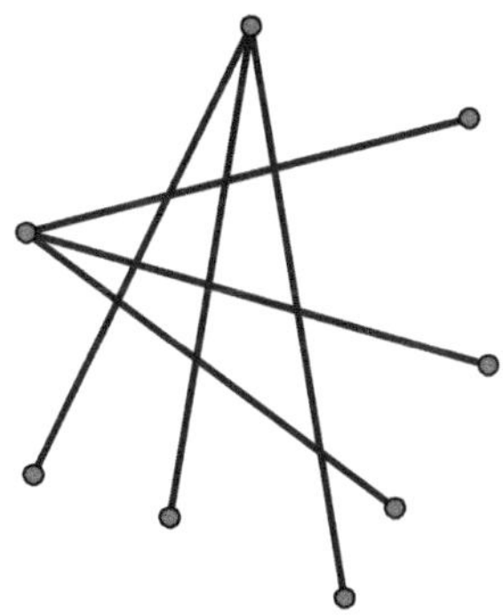

Figure 3-22

Organized Counting Is the Solution

In Figure 3-23, a square is separated into nine small squares, with diagonals drawn in each small square. The challenge here is to discover how many squares can be seen in the figure.

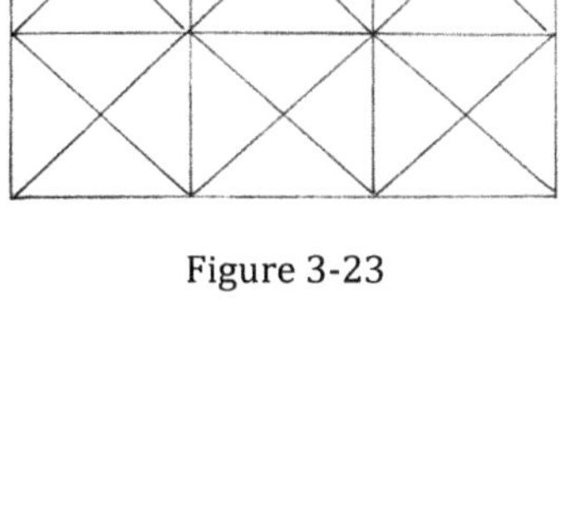

Figure 3-23

It would be an inconvenient task to just count squares that one sees regardless of the size and orientation. The sensible thing is to identify the size and orientation of each of the squares in the figure, as we show in Figure 3-24.

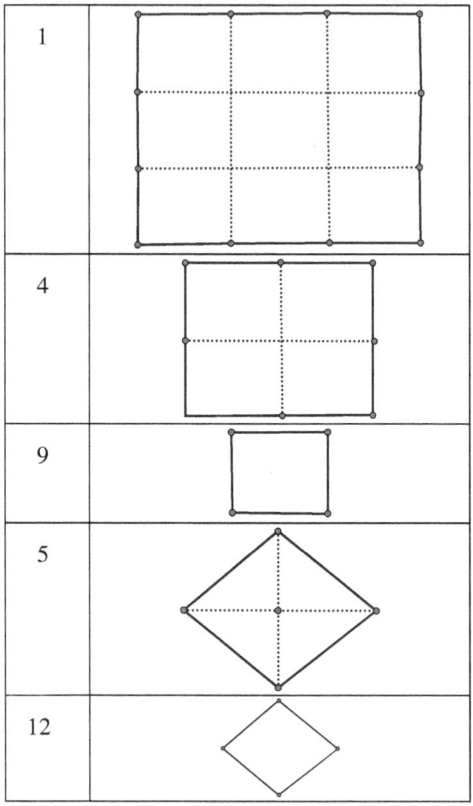

Figure 3-24

The Challenge of Lining Up Points

The challenge here is to move exactly one dot in Figure 3-25, so that all the dots will be in four rows of three.

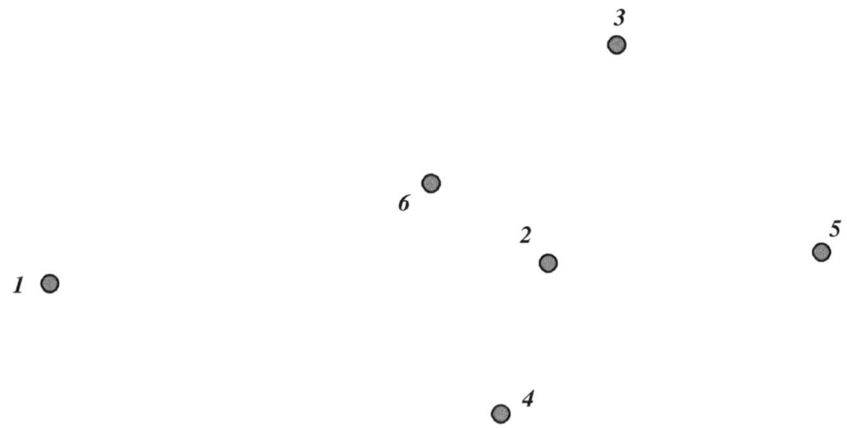

Figure 3-25

The trick to this challenge is to recognize the various rows of three that currently exist. Moving dot 2 would clearly ruin two extant rows, namely, 1-2-5 and 3-2-4. That would give you a clue that perhaps moving dot 6 would be a possible solution, as we show in Figure 3-26.

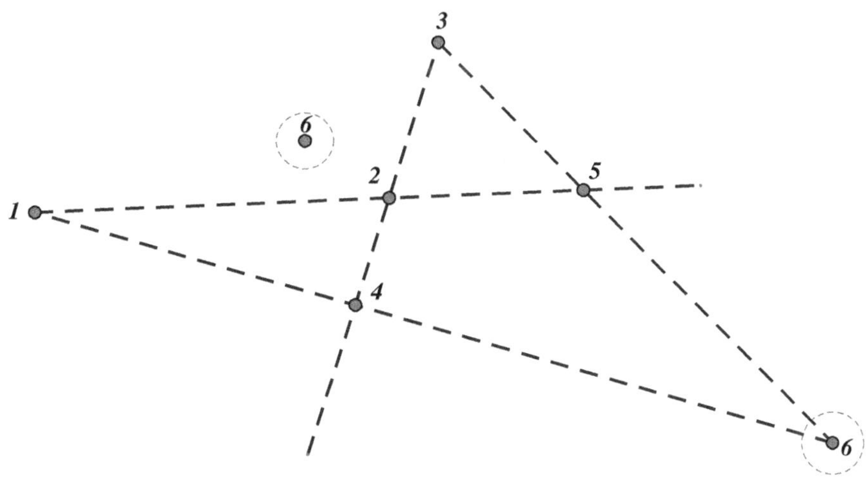

Figure 3-26

Correcting Matchstick Equations

Figure 3-27 presents a puzzle where matchsticks are shown with incorrect mathematics. The task is to move one matchstick to another position to make the mathematics correct.

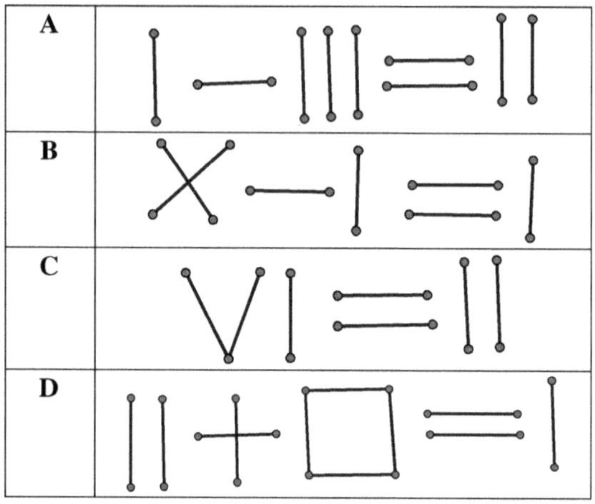

Figure 3-27

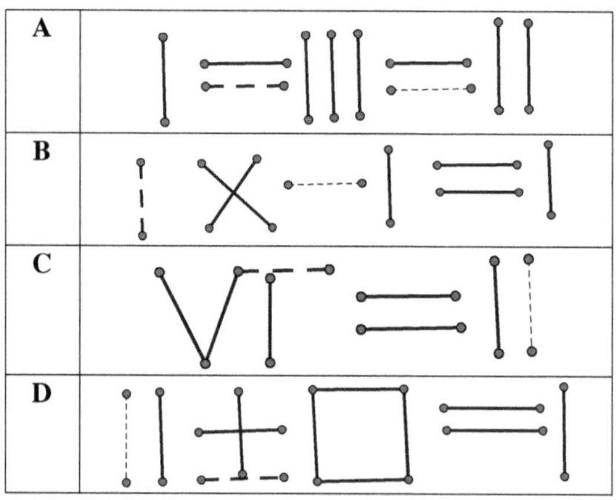

Figure 3-28

To make these matchstick equations correct, we need to move one stick to another position, as we show in Figure 3-28. Notice, in each case only one stick has been moved from the thin dashed line to the position in bold dashed line.

Cleverly Using Extremes

A car is driving along a highway at a constant speed of 55 miles per hour. The driver notices a second car exactly $\frac{1}{2}$ mile behind her. The second car passes the first exactly 1 minute later. How fast was the second car traveling, assuming its speed is constant?

According to many textbooks, the traditional solution is to set up a series of "Rate × Time = Distance" boxes, as in Figure 3-29.

Rate ×	Time =	Distance
55 mph	$\frac{1}{60}$ hr	$\frac{55}{60}$ miles
x mph	$\frac{1}{60}$ hr	$\frac{x}{60}$ miles

Figure 3-29

Then equating the distances, we get $\frac{55}{60}+\frac{1}{2}=\frac{x}{60}$, so that $x = 55 + 30 = 85$, which is the speed of the second car.

A more efficient method for solving this problem is to employ the strategy of *considering extremes*. We assume that the first car is going *extremely* slowly, that is, at 0 miles per hour. Under these conditions, the second car travels $\frac{1}{2}$ mile in one minute to catch the first car. Thus, the second car would have to travel $\frac{1}{2}\cdot 60 = 30$ miles per hour. When the first car is moving at 0 miles per hour, the second car travels 30 m.p.h. faster than the first car. If, on the other hand, the first car travels at 55 m.p.h., then the second car must be traveling at 85 miles per hour (within the legal limit, of course!).

Creatively Connecting the Dots

There are problems beyond mathematics that challenge our logical thinking. That is to say, the problem abuses our natural course of thought unless we consider the *extreme situations*. We offer one such problem here. Given nine dots as arranged in Figure 3-30, without lifting the pencil, use four straight lines to connect all nine dots.

Figure 3-30

When asked to draw four straight lines to connect the nine dots without lifting the pencil, most people would draw along the four sides of the square, only to realize that the center dot has been omitted. The next step would be to include the center dot and then find that the other dots have been omitted. For most people, there appears to be a psychological barrier to having the lines *extend* beyond the square; this is a sort of extreme. However, Figure 3-31 presents two variations of how this can be accomplished. Curiously, most people won't employ extremes and "think outside of the box!"

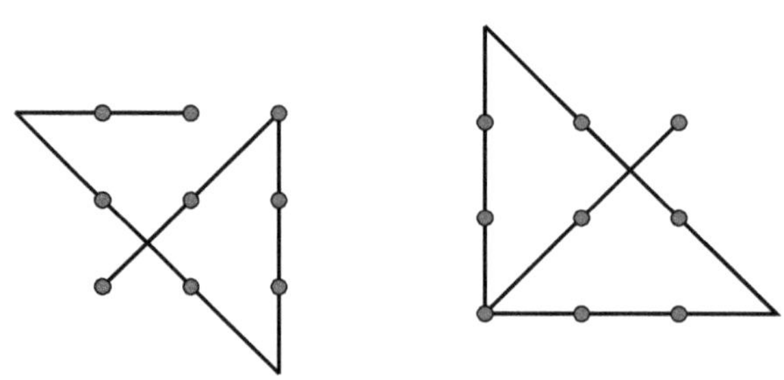

Figure 3-31

Thinking Cleverly

This problem demonstrates how going to the extreme solves a seemingly difficult problem immediately. Suppose we know that the number of water lilies in a pond doubles every day. After 100 days, the pond is completely covered with water lilies. The challenge is to determine how many days it will take for the pond to be half covered with water lilies. If we go to the *extreme*, namely, the last day, then we know that the water lilies have half-covered the pond, since on the last day they doubled their numbers to fully cover the pond on the hundredth day. Therefore, the lilies required 99 days to cover half the pond.

It's All Relative

The concept of relativity as developed by Albert Einstein is generally not well understood by most non-scientists, although it has many applications. It may be a difficult concept to grasp for some, so we shall exercise patience and support as we gently navigate further. We begin by presenting the following problem:

> *While rowing his boat upstream, Charles drops a cork overboard and continues rowing for 10 more minutes. He then turns around, chasing the cork, and retrieves it when the cork has traveled one mile downstream. What is the rate of the stream?*

The traditional method for solving this problem, a common one in algebra textbooks, is as follows. The water carries Charles 1 mile downstream, as well as carrying the cork 1 mile downstream. We let t equal the time the cork travels downstream after the initial 10 minutes, and have r equal the speed of rowing in still water, and s equal the rate of the stream. Recalling that the product of rate and time equals the distance traveled, we find the distance traveled by the cork to be $(10 + t)(s) = 1$ mile. Now the distance Charles, who has turned about and is following the cork, must travel is $t\ (r + s)$, which equals the return distance of his upstream trip, $10(r - s)$, plus the 1 mile that the cork travelled downstream. Therefore, $t = 10$. So, the time the cork had been traveling downstream is $10 + 10 = 20$ minutes, and 20 minutes to travel 1 mile translates to the stream flowing at a rate of 3 miles per hour.

Rather than approaching this problem through the traditional method as we did above, consider the following *extreme situation.*

The problem can be significantly simplified by considering the concept of relativity. It does not matter if the stream is moving and carrying Charles downstream, or if it is still. We are concerned only with the separation and coming together of Charles and the cork. Consider the *extreme situation* where the stream is stationary; Charles would spend as much time rowing to the cork as he did rowing away from it. That is, he would require $10 + 10 = 20$ minutes. Since the cork travels one mile during these 20 minutes, the stream's speed is 3 miles per hour.

The Classic Juice in the Bottle Problem

We can now consider a classic problem where the extremes method is essential. We have two 1-gallon bottles. One contains 1 quart of grape juice, and the other contains 1 quart of apple juice. We take 1 tablespoon of grape juice and pour it into the apple juice. Then, we take 1 tablespoon of this new mixture (apple juice and grape juice) and pour it into the bottle of grape juice. Is there more grape juice in the apple juice bottle or more apple juice in the grape juice bottle?

We can figure out this mixture problem in any of the usual ways, or we can be "clever" and use the strategy of *using extremes*.

To achieve this, we will consider the tablespoon quantity to be slightly larger. We will use an extremely large quantity. We'll let this quantity be the entire 1 quart. That is, following the instructions given in the problem, we will take the entire amount of grape juice and pour it into the apple juice bottle. This mixture is now 50% apple juice and 50% grape juice. We then pour 1 quart of this mixture back into the grape juice bottle. The mixture is now the same in both bottles. Therefore, there is as much apple juice in the grape juice bottle as there is grape juice in the apple juice bottle!

We can consider another *extreme case*, where the spoon used for transporting the juice cannot hold any quantity of liquid. In this case, the conclusion follows immediately: There is as much grape juice in the apple juice bottle as there is apple juice in the grape juice bottle, that is, zero.

Counting Games Problem

Here is a dramatic example of how extremes make solving a problem considerably simpler than the traditional method. A basketball tournament that has 25 teams competing. When a team loses a game, it is eliminated from the tournament. How many games must be played for there to be a single winner? The traditional approach would be to simulate the competition by having 12 teams play against another 12 teams, with a single team drawing a bye. Then the remaining teams would play against each other, with the points counted each time. However, using an *extreme situation*, we can imagine that one team is a professional basketball team and will never lose to another team. That implies that each of the teams will play against the professional team and lose. This will result in 24 games being played, with the professional team as the winner. Another way of thinking in extremes here is that, *finally*, 24 teams have to be removed from the tournament by losing before there is a single victor (not necessarily one professional team), yielding 24 necessary games.

An Algebraic Trick

Here is a rather harmless-looking problem that could be offered to any elementary algebra student. Yet, going to the extreme in this case suggests reaching the end of the problem, so that the challenge becomes trivial. We are given the following facts. The sum of two numbers is 2, and the product of the same two numbers is 3. We are asked to find the sum of the reciprocals of these two numbers.

The normal approach to this problem is to immediately set up the following two equations: $x + y = 2$, and $xy = 3$. Typical algebra training prepares us to solve these two equations simultaneously. That would have us change the first equation to read $y = 2 - x$, and then substitute this value for y into the second equation to get $x(2 - x) = 3$, which leads to the quadratic equation $x^2 - 2x + 3 = 0$. Using the quadratic formula[1] to solve this equation, we find that $x = 1 \pm i\sqrt{2}$. We would

---

[1] The quadratic formula for the equation $ax^2 + bx + c = 0$ is $x = \frac{-b \pm \sqrt{b^2 - 4ac}}{2a}$.

then need to find the value of y. We would have to take the reciprocals, and then add them to get the required answer, which is rather cumbersome and leads us to complex numbers. The curious part of this original problem is that it can be solved very simply if we focus on what we have been asked to find, rather than being distracted by finding the values of x and y. In other words, go *extremely* to the desired result.

We have been asked to find the sum of the reciprocals. That is, $\frac{1}{x}+\frac{1}{y}$. So let's find the sum of the reciprocals: $\frac{1}{x}+\frac{1}{y}=\frac{x+y}{xy}$, which essentially gives us the answer, since we know both the numerator and denominator from the given information: $x + y = 2$, and $xy = 3$. Hence, $\frac{1}{x}+\frac{1}{y}=\frac{x+y}{xy}=\frac{2}{3}$, and our problem is solved. Notice that by moving extremely to the end of the problem we achieved a very elegant solution that otherwise would have been rather complicated.

A Deceptive Trick

The challenge here is to take the cross that is shown on the right side of Figure 3-32 and place it on that array of dots so that five dots will be in the cross and eight dots will be outside the cross. This is a psychological trick, as you will see when you see the solution. But don't look below until you try to solve the problem yourself.

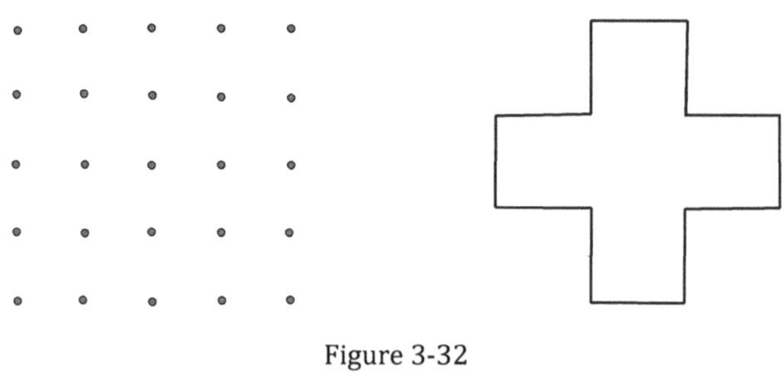

Figure 3-32

Typically, one maneuvers the cross along the various array of dots keeping the same figure orientation. That is where the trick comes in as you can see in Figure 3-33, where the cross is rotated 45° and then placed on the dots. Notice, five dots are within the cross and eight dots lie outside the cross.

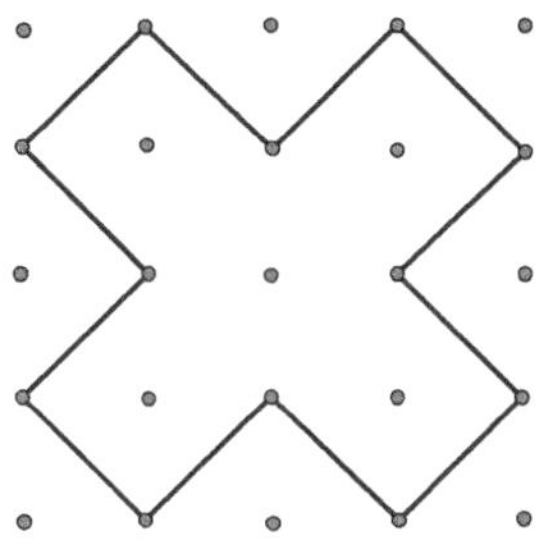

Figure 3-33

Creating a Basic Magic Circle

Figure 3-34 shows a circle with ten numbers sequentially indicated. The challenge here is to rearrange these numbers from 1 through 10 such that the sum of any two adjacent numbers is equal to the sum of any pair of adjacent numbers on the opposite ends of the diameters.

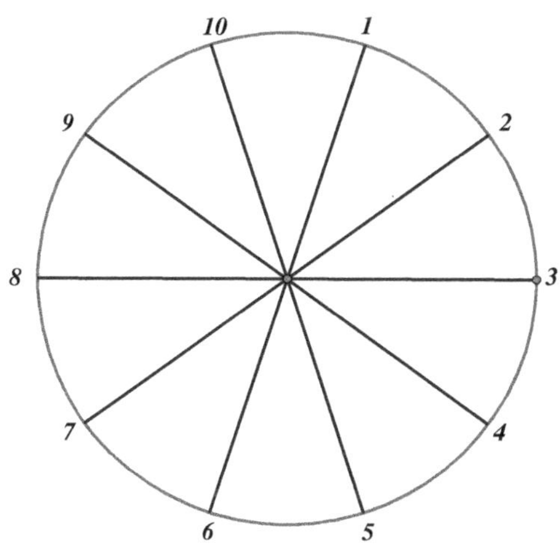

Figure 3-34

When one begins rearranging the numbers through trial-and-error such that the sum of the adjacent numbers on one end of the diameter is equal to the sum of the partner pair on the other end of the diameter, eventually a system will develop. In Figure 3-35, we show one possible solution.

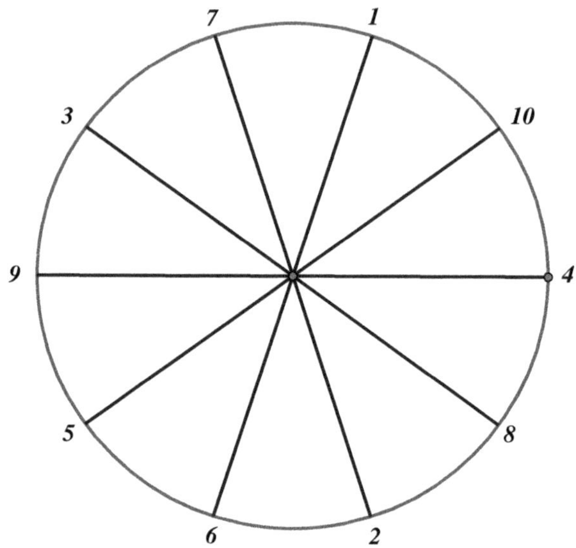

Figure 3-35

A Magic Circle

The arrangement of the numbers around the circle shown in Figure 3-36 has the following peculiar relationship: The sum of the squares of any two adjacent numbers is equal to the sum of the squares of the two numbers diametrically opposite them, as we show here:

$$8^2 + 14^2 = 2^2 + 16^2.$$

That is,

$$64 + 196 = 260 = 4 + 256$$

$$47^2 + 26^2 = 49^2 + 22^2.$$

That is,

$$2209 + 676 = 2885 = 2401 + 484$$

$$26^2 + 13^2 = 22^2 + 19^2.$$

That is,

$$676 + 169 = 845 = 484 + 361$$

$$2^2 + 49^2 = 47^2 + 14^2.$$

That is,

$$4 + 2401 = 2405 = 2209 + 196.$$

We may consider this a truly magic circle.

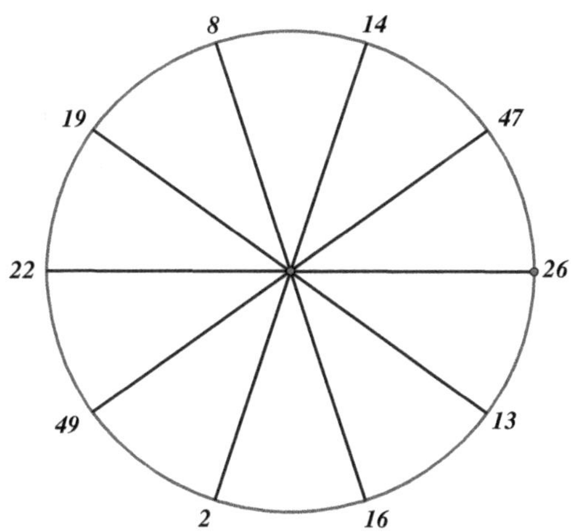

Figure 3-36

An Introduction to a Magic Triangle

We are familiar with magic squares, but here we extend the magic within mathematics to the triangle. The triangle pictured in Figure 3-37 contains three sections of numbers labeled A, B, and C. The magic manifests itself in the following way: Any sequence of numbers along a line from cell A to cell B or from cell A to cell C has a total of 380. Any sequence of numbers along a line from cell B to cell C also has a total of 380. Any sequence of five numbers in a straight line in each of the cells will have a total of 190.

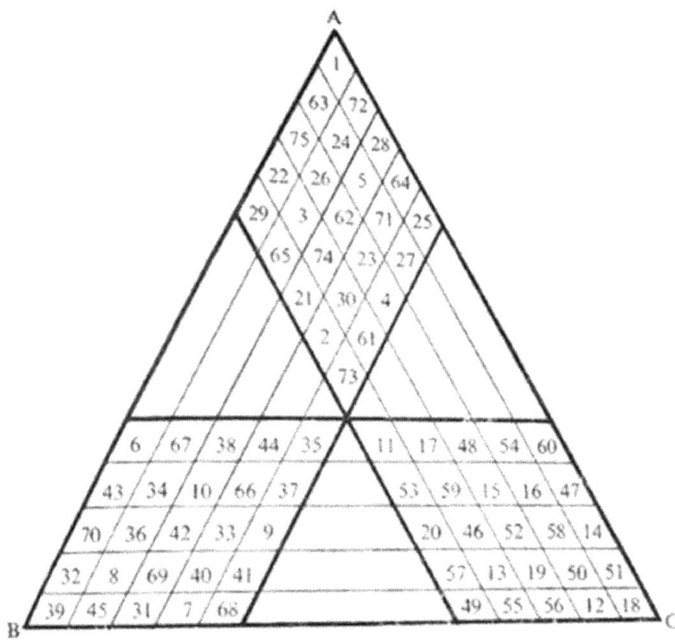

Figure 3-37

Some Logical Thinking

Suppose you have three balls of different weights, and you are told that the largest ball weighs as much as the other two balls. Furthermore, you are told the smallest ball weighs three pounds less than half the weight of the other two balls together. You are also given the fact that the three balls weigh 18 pounds in total. The challenge is to determine the weight of each of the three balls.

Logical thinking will let you conclude that the weights of the three balls are 9 pounds, 5 pounds, and 4 pounds. This can be justified by seeing that 9-pound ball has a weight of 5 + 4 pounds. The smallest ball weighs 4 pounds, as obtained by $\frac{1}{2}(9+5)-3=4$. And the total weight of the three balls is $9+5+4=18$ pounds.

A Curious Age-Old Puzzle

This is an age-old puzzle that has entertained countless people. If, on the average, a hen and a half can lay an egg and a half in a day and a half, how many eggs will six hens lay in eight days?

There are several ways to solve this problem. We begin with the following:

Since $\frac{3}{2}$ hens work for $\frac{3}{2}$ days, we can consider the job as $\left(\frac{3}{2}\right)\left(\frac{3}{2}\right)=\frac{9}{4}$ "hen-days". Similarly, the second job is $6\times 8=48$ "hen-days". Thus, we can create the following proportion, where x is the amount of eggs laid by 6 hens in 8 days:

$$\frac{\frac{9}{4}\text{ hen-days}}{48\text{ hen-days}}=\frac{\frac{3}{2}\text{ eggs}}{x\text{ eggs}}.$$

This yields the cross multiplication $\frac{9}{4}x=48\left(\frac{3}{2}\right)=72$. Therefore, $x=32$.

The alternative method of solution involves continuous doubling, where each line doubles the previous line:

1. $\frac{3}{2}$ hens lay $\frac{3}{2}$ eggs in $\frac{3}{2}$ days. Now we double the number of hens to get:
2. 3 hens lay 3 eggs in $\frac{3}{2}$ days. Now we double the number of days to get:

3. 3 hens lay 6 eggs in 3 days. Now we divide the number of days by 3 to get a one-day output:
4. 3 hens lay 2 eggs in 1 day. Now we double the number of hens:
5. 6 hens lay 4 eggs in 1 day. Now we need to multiply by the number of days sought (8):
6. *6 hens lay 32 eggs in 8 days.*

An Election Conundrum

To win an election, Barbara needs $\frac{3}{4}$ of the votes cast. After $\frac{2}{3}$ of the votes have been counted, Barbara has $\frac{5}{6}$ of what she needs. What part of the remaining votes does Barbara need to win?

The problem can be easily solved by using v to represent the total number of votes. After $\frac{2}{3}v$ have been counted, Barbara has $\frac{5}{6}$ of what she needs to win, which is $\frac{3}{4}v$. Therefore, she now has $\frac{5}{6}\left(\frac{3}{4}v\right)=\frac{5}{8}v$. Since she needs $\frac{3}{4}v$ to win, she still needs $\frac{6}{8}-\frac{5}{8}=\frac{1}{8}v$ to win. However, $\frac{1}{3}$ of the votes have not yet been counted. Hence, $\frac{\frac{1}{8}v}{\frac{1}{3}v}$ represents the part of the remaining votes which Barbara still needs. Simplifying this complex fraction, we find that Barbara needs $\frac{3}{8}$ of the remaining votes to win the election.

A Work Issue to Solve

Ernie can paint a certain job in 36 hours. After he has worked 20 hours, Alice joins him, and they finish the job after 10 more hours. If they had been working together the entire time, how long would it have taken Ernie and Alice to do the entire job?

This problem requires logical reasoning. Ernie works for 20 hours and completes $\frac{20}{36}=\frac{5}{9}$ of the painting job. Since the remaining $\frac{4}{9}$ of the painting job is done by both Ernie and Alice working together for 10 hours, we can set up the proportion $\frac{4}{9}=\frac{10}{x}$, where x represents the number of hours it will take the two of them working together complete the job, that is, $x=22\frac{1}{2}$ hours.

An Enigma to Solve

A hungry hunter met two farmers. One farmer had three loaves of bread, and the other farmer had five loaves of bread, all of the same size. The loaves were to be divided equally among the three men. The hunter paid \$8 for his share. How should the farmers divide the money?

We begin by considering the table in Figure 3-38:

	First Farmer	Second Farmer	Hunter
Amount given to the fund	3 Loaves	5 Loaves	\$8
Share taken from the fund	$\frac{8}{3}$ Loaves	$\frac{8}{3}$ Loaves	$\frac{8}{3}$ Loaves
Amount actually left in the fund	$\frac{1}{3}$ Loaf	$\frac{7}{3}$ Loaves	

Figure 3-38

Since the hunter pays \$8 for $\frac{8}{3}$ loaves, each loaf is worth \$3. Therefore, the first farmer receives $\frac{1}{3}$ of $\$3 = \$1$ for his donation, while the second farmer receives $\frac{7}{3}$ of $\$3 = \$7$ for his donation.

Locating Squares

Figure 3-39 shows an arrangement of 12 dots. The task is to determine how many squares can be found so that each vertex of each square is on a dot.

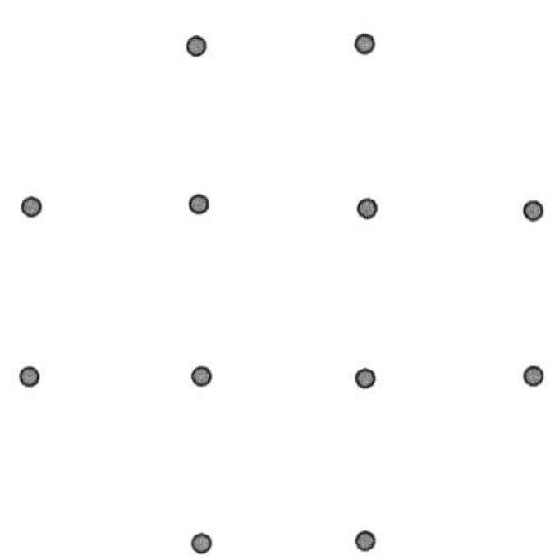

Figure 3-39

When we connect the dots as shown in Figure 3-40, we find that there are five small squares

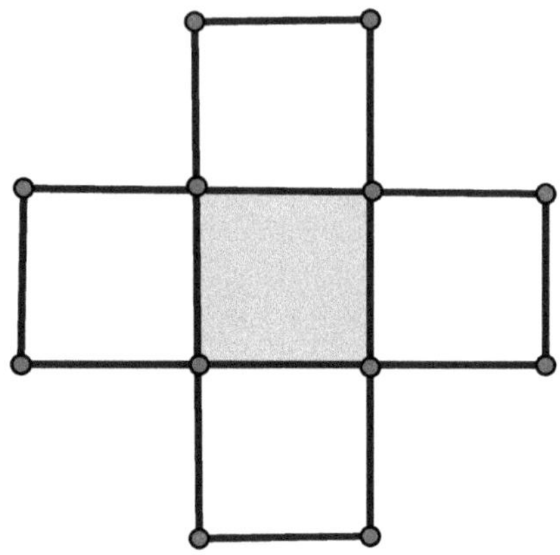

Figure 3-40

Connecting the dots as we have in Figure 3-41, we find that there are four medium size squares.

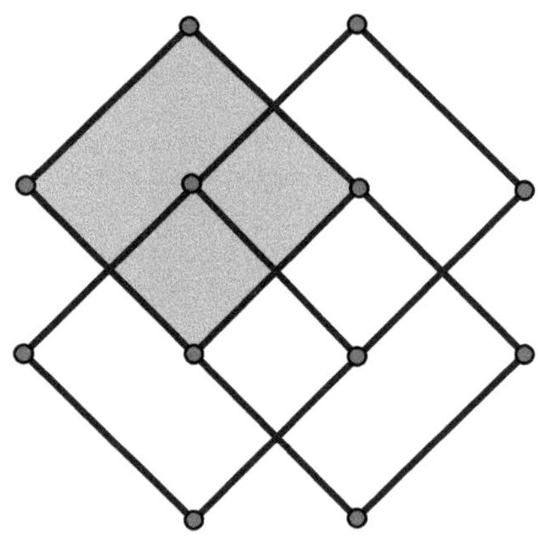

Figure 3-41

Finally, when we connect the dots as shown in Figure 3-42, we find that there are two larger squares. All in all, there are 11 squares visible in this arrangement where every square has its vertices on a dot.

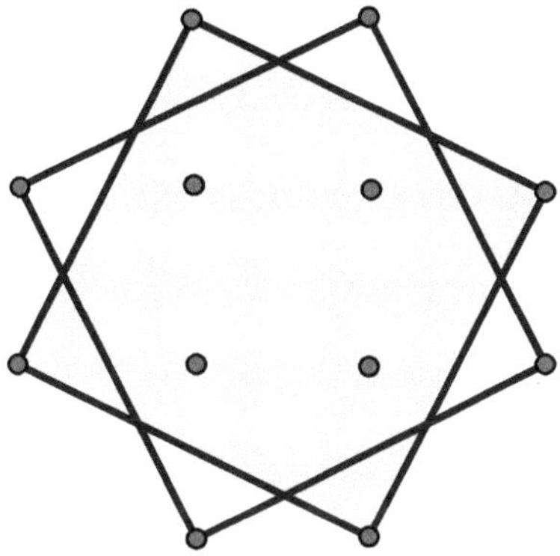

Figure 3-42

A Worthless Increase

Suppose you had a job where you received a 10% raise. Because business is falling off, the boss is soon forced to give you a 10% cut in salary. Will you be back to your starting salary? The answer is a resounding (and very surprising) NO!

This little story is quite disconcerting, since one would expect that with the same percentage increase and decrease you should be where you started from. This is intuitive thinking, but wrong! Convince yourself of this by choosing a specific amount of money and following the instructions.

Begin with $100. Calculate a 10% increase on the $100 to get $110. Now take a 10% decrease of this $110 to get $99, or $1 less than the starting salary.

You may wonder where the result would have been different if we first calculated the 10% decrease and then the 10% increase. Using the same $100 basis, we first calculate a 10% decrease to get $90. Then the 10% increase yields $99, the same as before. Clearly, order makes no difference.

A similar situation, that is, one that is deceptively misleading, can be faced by a gambler. Consider the following situation. You may even simulate it with a friend to see if your intuition bears out.

You are offered a chance to play a game. The rules are simple. There are 100 cards, face down. 55 of the cards say "*win*" and 45 of the cards say "*lose*." You begin with a bankroll of \$10,000. You must bet one-half of your money on each card turned over, and you either win or lose that amount based on what the card says. At the end of the game, all cards have been turned over. How much money do you have at the end of the game?

The same principle as above applies here. It is obvious that you will win ten times more than you will lose, so it appears that you will end with more than \$10,000. What is obvious is often wrong, and this is a good example of this principle. Let's say that if you win on the first card, you now have \$15,000. Now you lose on the second card; you now have \$7,500. If you had first lost and then won, you would still have \$7,500. So, every time you win one game and lose one game, you lose one-fourth of your money. Thus, you end up with $10{,}000\cdot\left(\frac{3}{4}\right)^{45}\cdot\left(\frac{3}{2}\right)^{10}$. This is \$1.38 when rounded off. Surprised?

Weighing Coins Enigma

Suppose you have nine coins that look identical but one coin is lighter than the other eight coins. The task is to locate the lighter coin, if you are allowed only two weighings on a balance scale.

The solution to this task is quite remarkable. Randomly select two groups of three coins from the total of nine coins. Place the two batches of coins on the scale. If the scale balances, the lighter coin is among the three excluded coins. Take this last group of three and select two coins to be weighed against each other. If they don't balance, then you have identified the lighter coin. If they do balance, the third coin from the last group is the lighter coin.

If, on the first weighing, the scale does not balance, then the group of three which is lighter includes the lighter coin. You then revert to the process described above, weighing two coins from the group of three to determine which is the lighter one. If they balance, then, as we indicated above, the lighter coin is the one which you haven't yet had to weigh.

The Famous Train Problem

This is a question that has been facing the general public for years. If the train whose length is 1 mile is traveling at a rate of 60 m.p.h., how long will it take the train to pass through a 1-mile-long tunnel?

As the train enters the tunnel, it must travel two miles before the last car clears the tunnel. At 60 miles an hour, the train travels 1 mile every minute. Therefore, it will take 2 minutes to clear the tunnel completely.

A Family Question

There are times when we seek the number of siblings in a family. How many brothers and sisters are there in a family in which each boy has as many sisters as he has brothers and each girl has twice as many brothers as she has sisters?

Using logical reasoning to solve the problem begins this way. If each boy has an equal number of brothers and sisters, then there must be one boy more than the number of girls in the family. Suppose there are two boys and one girl, or three boys and two girls, or four boys and three girls. The latter case meets the requirement that each girl has twice as many brothers and sisters. Trial-and-error can be a useful technique, as we see here.

The 10-Coin Challenge

The challenge is to place ten coins in such a way that they lie in five straight lines, where each line has exactly four coins. There are numerous solutions, and we offer two solutions in Figure 3-43.

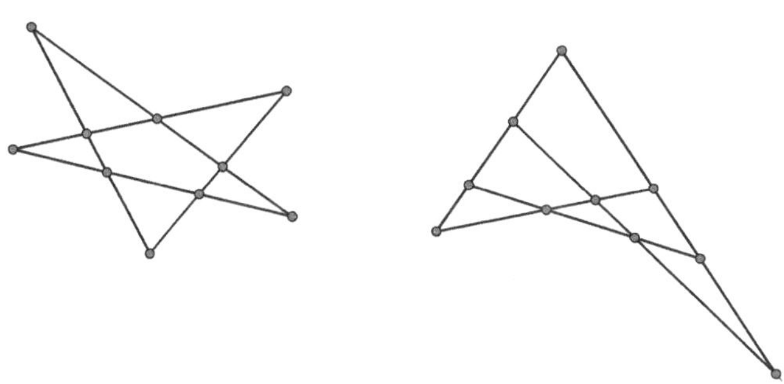

Figure 3-43

The 28-Coin Challenge

The challenge is to place 28 coins in such a way that they lie in twelve rows with exactly seven coins in each row.

Figure 3-44 presents one possible solution. The logical approach could be to take the two lines *AB* and *AC*, place seven coins on each line, and then join the points sequentially, as shown, where each intersection is a location of the coins.

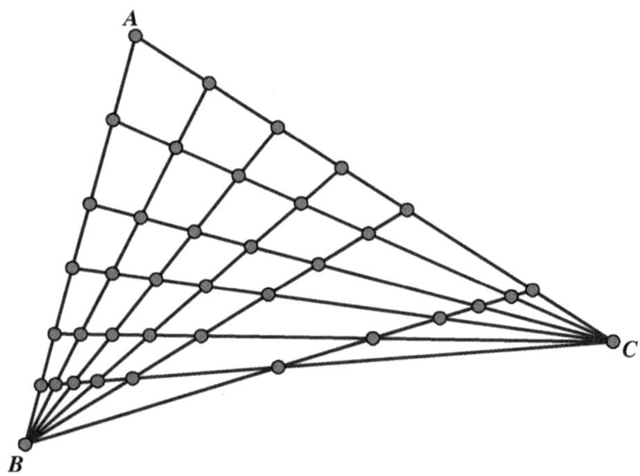

Figure 3-44

A Pentagram Challenge

The pentagram in Figure 3-45 shows sixteen dots arranged in fifteen lines consisting of four dots. The challenge here is to rearrange the points so that they form ten lines of four dots with no two lines parallel. Try it before looking further.

Since pictures speak louder than words, we show in Figure 3-46 the ten lines consisting of four dots each made from the original configuration. Problem solved!

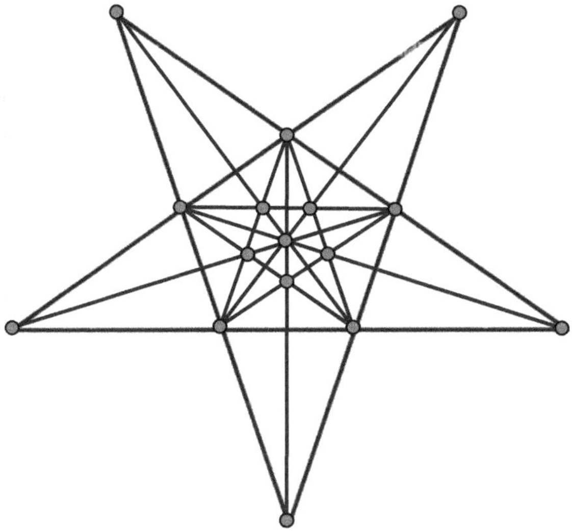

Figure 3-45

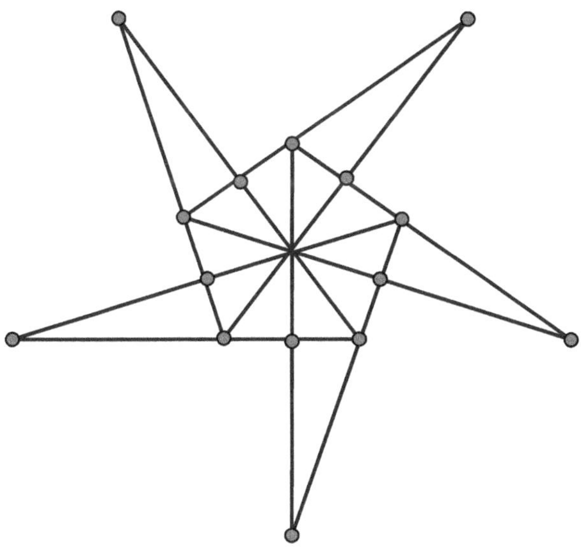

Figure 3-46

Clever Thinking

If a bag of sand balances on a scale with $\frac{3}{4}$ of a bag of sand and a $\frac{3}{4}$ pound weight, what is the weight of a full bag of sand?

If a full bag of sand balances with $\frac{3}{4}$ of a bag of sand and a $\frac{3}{4}$ pound weight, then logical thinking dictates that $\frac{1}{4}$ of the bag of sand must weigh as much as the $\frac{3}{4}$ pound weight. That is, $\frac{1}{4}$ bag of sand $= \frac{3}{4}$, or the bag of sand weighs 3 pounds.

Comparing Candle Lengths

Suppose we have two candles of equal length that are lit at the same time. One of the candles burns down completely in six hours and the other candle burns down in three hours. After how much time, while the candles are burning, will the slower-burning candle be exactly twice as long as the faster-burning candle?

After two hours of burning, the slower candle will be twice as long as the faster candle. At that particular time, the slower candle will have $\frac{2}{3}$ of its length left while the other candle will have only $\frac{1}{3}$ of its length left. In other words, after two hours, the slower-burning candle will be twice as long as the faster-burning candle.

Emptying a Swimming Pool

This question has often challenged swimming pool owners. If they want to empty the pool, are they better off with one huge hose 8 inches in diameter or two small hoses with 4 inch diameters?

Most people think that two hoses are better than one, whereas some people feel that it makes no difference, as either one is okay. The geometric truth is that the only thing that matters is the circular cross-section area of each hose, measured by πr^2. With a diameter of 8 inches and a radius of 4 inches, the circular cross-section of the larger hose is 16π. Each smaller hose has a radius of 2 inches and a circular cross-section area of 4π. The total cross-section area of the two smaller hoses is 8π. We can exhibit the difference visually in Figure 3-47.

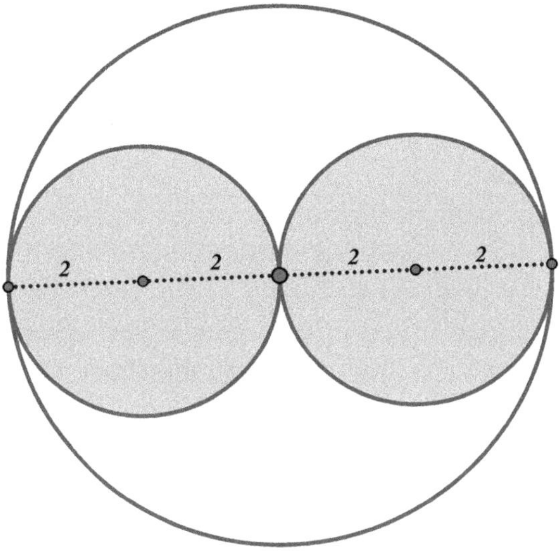

Figure 3-47

Meeting Trains Conundrum

Here is a time-tested puzzle that supersedes what students tend to learn in school because it's obvious. Consider two trains 850 miles apart. One train travels at the speed of 120 m.p.h. and the other train travels at 80 m.p.h. The two trains are traveling directly towards each other, and we need to determine how far apart they will be exactly one hour before they meet.

One hour before they meet the trains meet, they must be 120 miles + 80 miles = 200 miles. Note that their initial distance apart has no bearing on the answer.

Determining Marble Weights

Suppose you have two light marbles and two heavier marbles which are visually indistinguishable. The challenge is to determine which pair of marbles are the heavier marbles with two weighings on a balance scale.

We begin by weighing any two marbles against each other on the balance scale. If the scale does not balance, then we have identified one light marble and one heavy marble. We would then weigh the other two marbles and identify which one is the light and which one is the heavy marble.

On the other hand, if the scale balances when first weighing two randomly selected marbles, they must either both be heavy or both be light. This can be resolved by taking one of these balanced marbles and weighing it against one of the remaining marbles to determine whether it is a light or heavy marble. You then have sorted the marbles with two weighings.

Grouping Marbles

Suppose you have three red marbles, three white marbles, and three blue marbles. You want to know how many groups of three marbles, each having a red, a white, and a blue marble, you can make from these nine marbles.

This is simply a case of clever counting. Suppose we start with the three red marbles and three white marbles. We could make $3 \times 3 = 9$ groups with these red and white marbles. If each of these groups would get a blue marble, that would add another marble to each of the groups, so that we then have $3 \times 9 = 27$ tricolor groups of marbles.

A Time Challenge on a Clock

You possess two clocks that have time problems. Consider when they both read 12 o'clock. From then on, one clock gains 1 minute each hour while the other clock loses 1 minute each hour. In how many hours will both clocks again show the identical time, namely, 12 o'clock?

Begin by noting that every hour the clocks are 2 minutes further apart in time. Therefore, after 30 hours they will be one hour apart in time. After 12 hours, properly functioning clocks would on be showing the same time. Therefore, $30 \times 12 = 360$ hours will be required to have them show the same time as they started with: 12 o'clock.

Place Coins Efficiently in Cells

The challenge provided here is to move the coins marked "heads" and those marked "tails" into the cells marked for the others, as shown in Figure 3-48. That is, to move the heads to where the tails are and vice-versa with the minimum number of moves. There is a restriction: the coins cannot be flipped over, but they can be slid into empty cells diagonally, vertically, or horizontally.

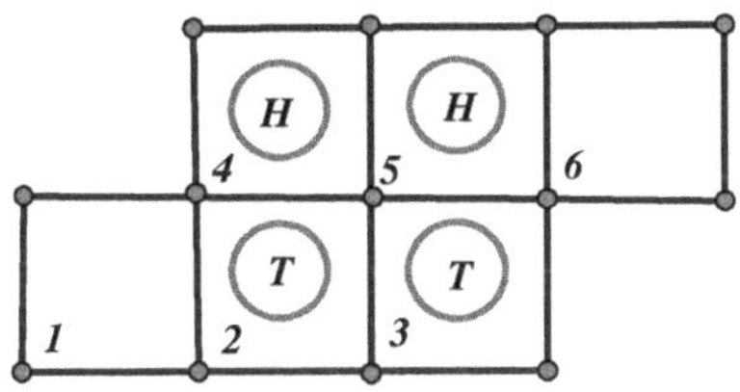

Figure 3-48

We will now demonstrate how this can be done in the least number of moves, which will be five moves. Begin by moving the coin from cell 5 to cell 6. (See Figure 3-49.)

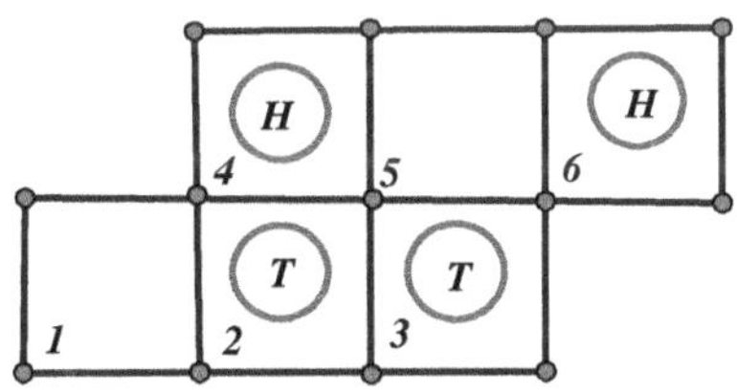

Figure 3-49

Then slide the coin from cell 2 to cell 5. (See Figure 3-50.)

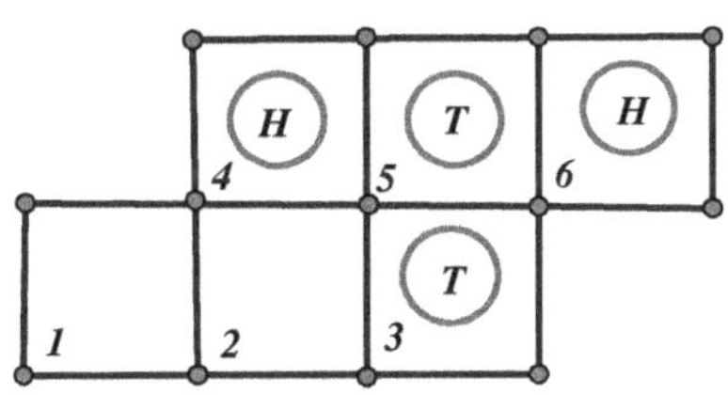

Figure 3-50

Then slide the coin from cell 4 to cell 2. (See Figure 3-51.)

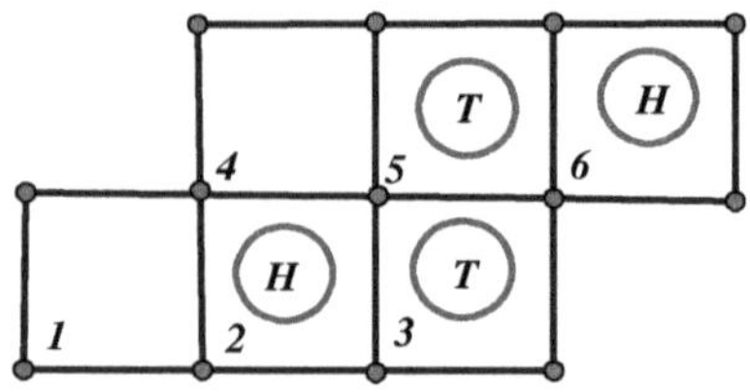

Figure 3-51

Next, slide the coin from cell 3 to cell 4. (See Figure 3-52.)

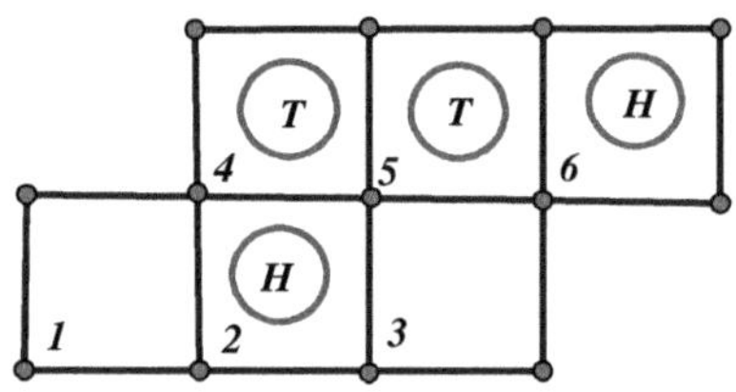

Figure 3-52

And lastly, we slide the coin from cell 6 to cell 3. (See Figure 3-53.)

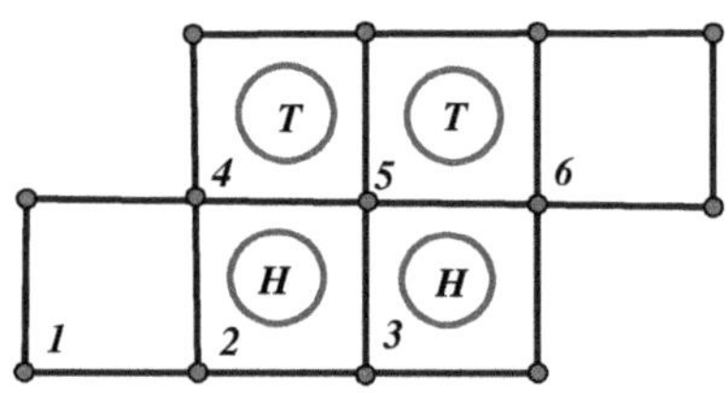

Figure 3-53

And so, we have moved the heads to where the tails were and vice-versa in five steps, which is the least number of moves required to make these changes under the conditions presented.

The Trick for Partitioning a Circle

The challenge here is to partition a circle into 22 sections using only six lines, as you can see in Figure 3-54. In order to succeed with this challenge, one must be sure that every line intersects every other line and no two intersections should share a point; that is, that every line passes through every other line once.

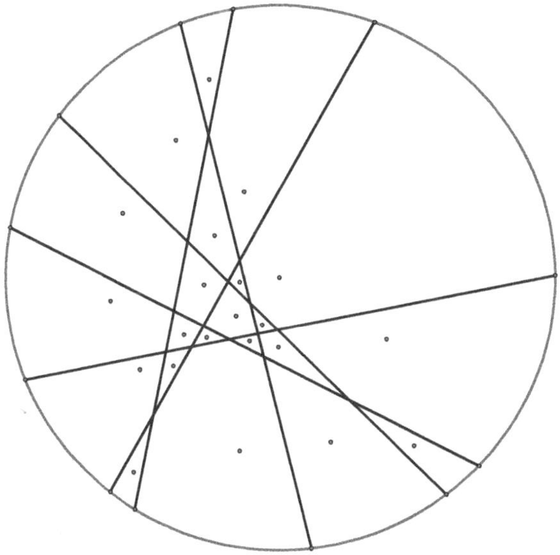

Figure 3-54

Guess the Ages

Following are a series of age-related problems along with their solutions. The reader would be wise to try and solve the problems before reading the answers.

> ***I.*** *A man does not want to reveal his age, but he does give the clue that his age and his wife's age have the digits reversed. He is clearly the elder of the two, and the difference between their ages is $\frac{1}{11}$ of the sum of their ages.*

Perhaps through trial and error a clever reader will determine that the husband's age is 54 and the wife's age is 45, as the difference of the ages is 9, which is $9=\frac{1}{11}(54+45)=\frac{1}{11}(99)$. You might wish to challenge your friends with this question.

II. A twelve-year-old son wants to take driving lessons. His father says to the son, "You can begin driving when I am three times as old as you." The father is currently 45 years old. At what age will his son be allowed to begin driving lessons?

In 4.5 years, the father will be 49.5 years old and in 4.5 years his son will be 16.5 years old, which is one third of his father's age. Therefore, when he reaches the age of 16.5 years, the son will be able to begin driving lessons.

III. A son asks his mother how old she is. Instead of giving him the answer, she makes a mathematics excursion for her son. She responds, "My age, your father's age, and your age have a sum of 70 years." The son then asks the father how old he is. The father responds, "I am currently six times as old as you are my son." The son then responds with a question: "Will I ever be half as old as you are now, father?" The father then responds affirmatively: "When that happens, our ages will have a sum exactly twice as much as their sum is today." If the son had been some years older, he might have been able to figure out the ages of his parents. The question is, what is the age of the mother?

After probing several possibilities, one could come up with the fact that the mother must be 29 years and 2 months old. The father's age would be 35, and the son's age would be 5 years and 10 months. Therefore, the sum of the ages would have been 70 years. The father's age is six times the son's age. After 23 years and 4 months, the sum of their ages would be 140 years, and the son will be 29 years and 2 months old.

IV. Lisa says to her brother David, "In seven years our combined ages will be 63." David responds, "When you were my current age, you were twice as old as I was then." What are the ages of David and Lisa?

After testing various options, you might have realized that Lisa's age must be $29\frac{2}{5}$ years, and David's age is $19\frac{3}{5}$ years. When Lisa was age $19\frac{3}{5}$, David was only $9\frac{4}{5}$. Therefore, Lisa was at that time twice as old as David.

V. A family has fifteen children, all born at intervals of 1 ½ years apart. The oldest daughter did not want to reveal her age to the census taker, but admitted she was just seven times older than the youngest sibling. What is the age of the oldest child?

The eldest child must be 24 years old, and her little sibling must have been 3 years old with 13 siblings between them. The trick is where the oldest sibling said that she was 7 times older than the youngest brother, which is equivalent to saying that she was 8 times older than the youngest sibling. Caution must be taken to avoid the trick and assume that the oldest daughter was 7 times as old as the youngest sibling, which would mean their ages are 24.5 and 3.5, which is incorrect.

VI. Max and Jack are brothers and the sum of their ages is 44 years. Max is twice as old as Jack was when Max was half as old as Jack will be when Jack is three times as old as Max when Max was three times as old is Jack. Our challenge is to find out how old Max is now.

We might be able to see that the ratio of Max's age to Jack's age is 5:3. Since the sum of the ages is 44 we can conclude that Max is $27\frac{1}{2}$ years old and Jack is $16\frac{1}{2}$ years old. The difference of their ages is exactly 11 years.

We will now insert these ages into our puzzle and see how it works out. Max $\left(27\frac{1}{2}\right)$ is twice as old as Jack $\left(13\frac{3}{4}\right)$ was when Max was half as old $\left(24\frac{3}{4}\right)$ as Jack will be $\left(49\frac{1}{2}\right)$ when Jack is three times as old $\left(49\frac{1}{2}\right)$ as Max was $\left(16\frac{1}{2}\right)$ when Max was $\left(16\frac{1}{2}\right)$ three times as old as Jack $\left(5\frac{1}{2}\right)$.

We can now check in the reverse. When Max was three times as old as Jack, Max was $\left(16\frac{1}{2}\right)$ and Jack was $\left(5\frac{1}{2}\right)$, that is, 11 years younger. Jack's age will be $\left(49\frac{1}{2}\right)$ when he is three times as old as Max was then. When Max was half this age he was $\left(24\frac{3}{4}\right)$. At that time, Jack must've been 11 years younger, or $\left(13\frac{3}{4}\right)$. Therefore, Max is now twice as old, namely, $27\frac{1}{2}$, and Jack, who is 11 years younger must be $\left(16\frac{1}{2}\right)$. The trick is to stay focused on the two ages as they progress.

A Trick to Construct Equal Areas

In Figure 3-55, we have six spaces of equal area formed by thirteen equal-length fences: three on the top, three on the bottom, and seven vertical fences. If we remove one of these fences, how can we create an arrangement that would provide six equal areas?

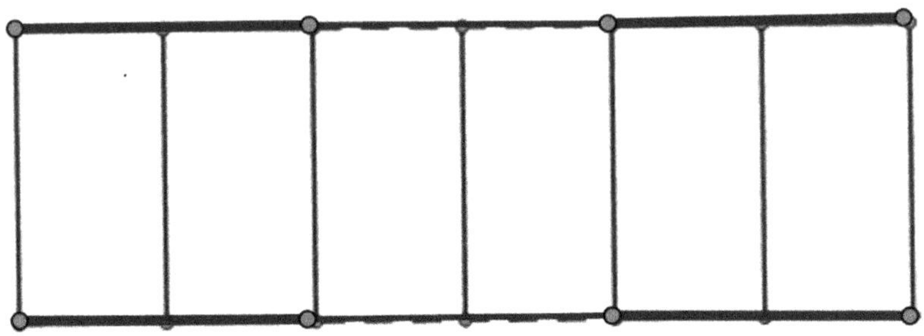

Figure 3-55

Removing one of these fence pieces leaves us with twelve pieces of fence. That should be a clue that we could create a regular hexagon where all the sides and diagonals are of equal length, which gives us the hexagon shown in Figure 3-56.

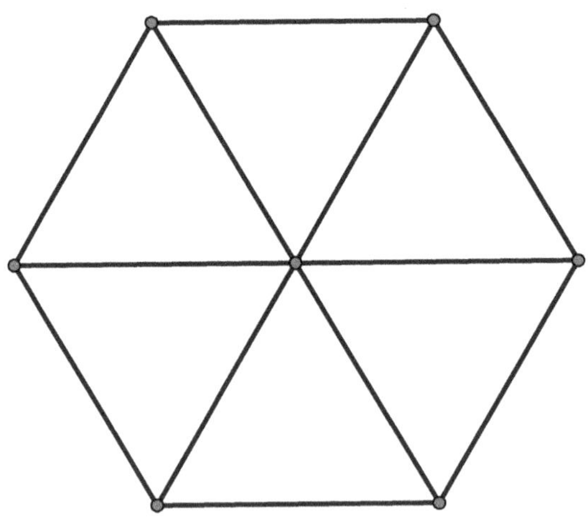

Figure 3-56

Notice, in both cases there are six equal area sections.

Puzzling Paradoxes

Paradoxes are fun to observe and yet have a very important message embedded within them. There is much to be learned through this entertainment. Here are some paradoxes that will give you something to think about.

We know that we can multiply by equals and the answers are appropriately correct. However, consider the following:

2 pounds = 32 ounces, and also ½ pound = 8 ounces.

By multiplying these above two equalities:

(2 × ½) pound = (32 × 8) ounces, which leads to an absurd result, that is, 1 pound = 256 ounces!

This paradox lies in the fact that the units were not treated properly, and can be best answered by considering the following example:

2 feet = 24 inches and also ½ foot = 6 inches.

By multiplying these equalities, we get 1 square foot = 144 square inches, which is correct.

Another paradox can be seen with the following:

We have $1 \times 0 = 2 \times 0$, and we know that $0 = 0$. By dividing these equalities, we arrive at the absurd result that $1 = 2$. Here, of course, we see the familiar rule against dividing by zero being broken, thus leading us to an absurd result.

The messages of each of these paradoxes should be clear: Don't just dive into an arithmetic process without thinking about what you are doing.

A Prime Number Magic Square

Recall that a magic square is a square arrangement of numbers where the sums of the rows, columns, and diagonals are the same. We have a special magic square which we will admire here. A *prime* magic square is one in which all the elements are prime numbers. The smallest sum of rows, columns, and diagonals of such a 3×3 magic square is 177, which you can verify with the magic square shown in Figure 3-57.

17	89	71
113	59	5
47	29	101

Figure 3-57

However, while on this special magic square, there is also something special about the sum generated. The number 177 is the sum of all the prime numbers from 2 to 47, that is,

$$2 + 3 + 5 + 7 + 11 + 13 + 17 + 19 + 23 + 29 + 31 + 37 + 41 + 43 + 47 = 177.$$

Incidentally, the fifteenth (1 + 7 + 7) prime number is 47, which was the last prime we used in this sum. How marvelous are the coincidences that can be discovered in mathematics!

Comparing Married Couples

In a town with 1000 married couples, $\frac{2}{3}$ of the husbands who are taller than their wives are heavier than their wives. Also, $\frac{3}{4}$ of the husbands who are heavier than their wives are also taller. If there are 120 wives who are taller and heavier than their husbands, how many husbands are taller and heavier than their wives?

We use simple algebra to solve this problem. We let x equal the number of husbands who are taller than their wives, let y equal the number of husbands heavier than their wives, and let z equal the number of husbands taller and heavier than their wives.

We then can get the following relationships directly from the statements: $z = \frac{2}{3}x$, and $z = \frac{3}{4}y$. From the 1000 husbands in the town, we have $1000 - x - y + z = 120$. From the first two equations we get $\frac{2}{3}x = \frac{3}{4}y$, which is $8x = 9y$. Simple algebra using the equation above yields $x = 720$ and $y = 640$. Therefore, the number of husbands who are taller and heavier than their wives is 480.

A Simple Confusing Question 1

Consider the following. You have a ball that weighs $\frac{9}{10}$ of its weight plus $\frac{9}{10}$ of a pound. What is the weight of the ball?

This sounds both simple and somewhat confusing, yet requires only elementary algebra. You must understand the wording to solve the problem. If we let x equal the weight of the ball, we can create and solve the following equation:

$$\frac{9}{10}x+\frac{9}{10}=x$$

$$9x+9=10x$$

$$x=9.$$

Therefore, the weight of the ball is 9 pounds.

Having now solved the problem algebraically, consider an analogous problem solved logically, as follows.

A Simple Confusing Question 2

On a balance scale we find that a certain brick can balance evenly against $\frac{3}{4}$ of a brick plus $\frac{3}{4}$ of a pound. The challenge is to find the weight of the entire brick.

The solution can be somewhat tricky. We have $\frac{3}{4}$ of a brick and "something else" on the right side of the scale, which is equal to the entire brick on the left side of the scale. That indicates that the "something else" must be the weight of $\frac{1}{4}$ of a brick. Therefore, $\frac{3}{4}$ of a pound is $\frac{1}{4}$ of the brick. Thus, the brick must weigh four times that, or $4\times\frac{3}{4}=3$ pounds.

The Counterintuitive Trick

Some logic tricks seem impossible. We are about to embark on one such trick now, one that has been played in restaurants and bars using toothpicks or matchsticks. Suppose you are presented with a collection of toothpicks arranged as shown in Figure 3-58, where each of the two rows and two columns contains 11 toothpicks.

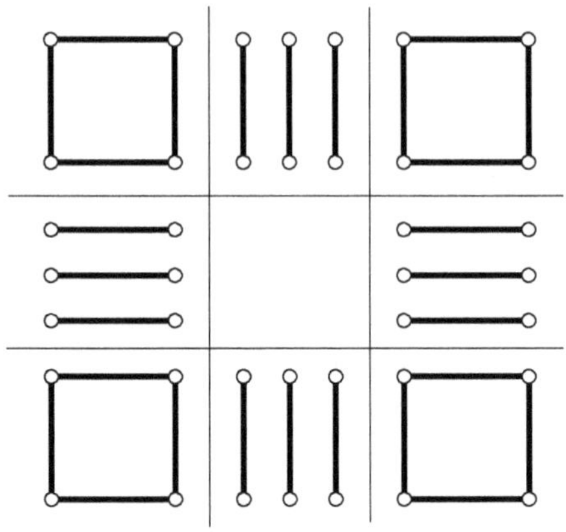

Figure 3-58

You are asked to remove one toothpick from each row and column and still keep 11 toothpicks in each row and column. This seems to be impossible, since we are actually *removing* toothpicks. Initially, this could perhaps look like the arrangement shown in Figure 3-59.

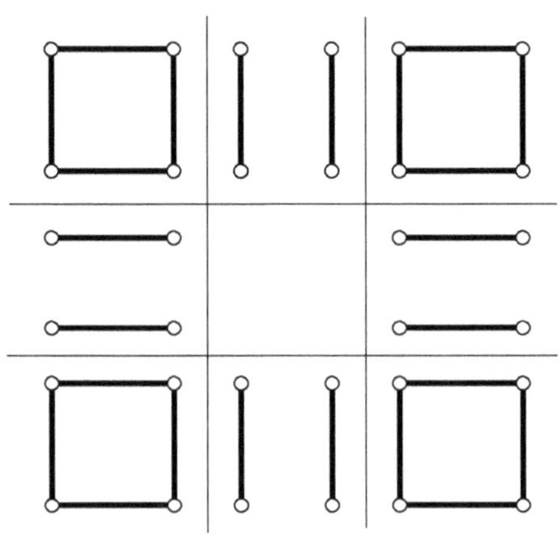

Figure 3-59

How can this possibly be done? After some contemplation, you will begin to see that this can be done if you count some toothpicks twice. In Figure 3-60, we have taken a toothpick from the center portion of each of the rows and columns and placed these toothpicks in the corners so that they can be counted more than once. This is the crux of the trick!

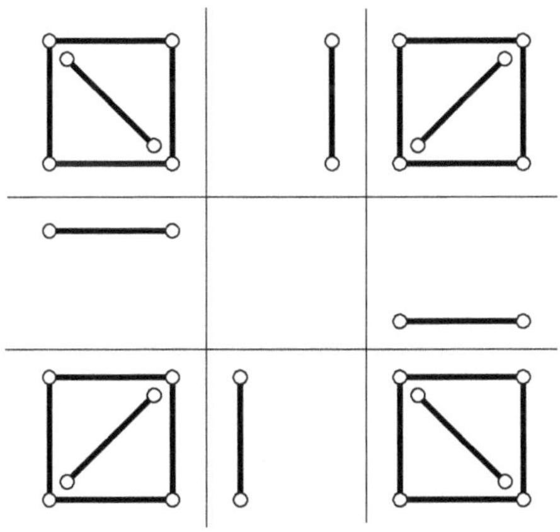

Figure 3-60

Thus, we have achieved our goal of having 11 toothpicks in each of the two rows and each of the two columns. This is a skill that merits attention, so that one can go through life by analyzing situations in a more critical fashion.

A Confusing Purchase Problem

Suppose you are buying 17 balls for $72. You pay $1 more for each of the red balls than for each of the blue balls. How many of each kind of ball have you bought?

Perhaps easiest way to approach this problem is using elementary algebra, where we let x be the number of blue balls bought each at y dollars. Then $17 - x$ must be the number of red balls bought at $y + 1$ dollars each. We can then set up the equation

$$\begin{aligned} xy+(17-x)(y+1)&=72 \\ xy+17y-xy+17-x&=72 \\ 17y&=55+x \\ y&=3+\frac{4+x}{17}. \end{aligned}$$

Now we have to do a little bit of analysis. The only value for x less than 17 that allows y to be an integer is $x = 13$. In that case, $y = 4$. We have purchased 13 blue balls at \$4 each and 4 red balls at \$5 each.

The Broken Chain Repair Problem

This trick can be done by considering four pieces of chain, each of which consists of three links, as shown in Figure 3-61. The trick here is to show how these four pieces of chain could be made into a circular chain by opening and closing *no more than* three links.

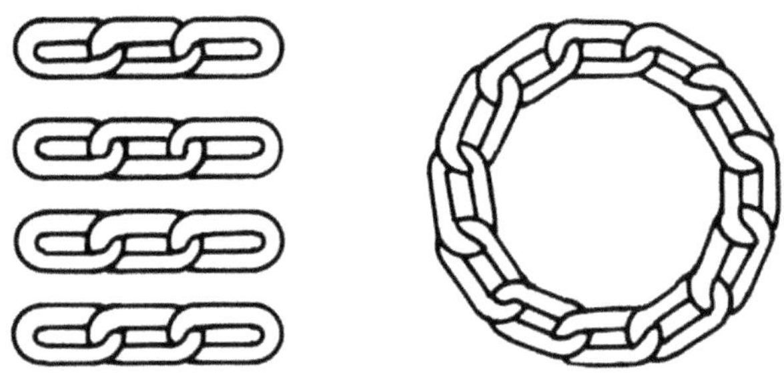

Figure 3-61

Your first attempt at a solution might involve opening the end link of one chain, joining it to the second chain to form a six-link chain; then opening and closing a link in the third chain and joining it to the six-link chain to form a nine-link chain. By opening and closing a link in the fourth chain and joining it to the nine-link chain, you obtain a twelve-link chain, but it is *not* a circular chain. Therefore, this attempt ends unsuccessfully. Most attempts usually involve other combinations of opening/closing one link of each of various chains pieces to join them together and get the desired result, but these approaches will not work.

Let's look at this from another point of view. Instead of continually trying to open and close *one* link of each chain piece, a different point of view would involve opening *all* three links in one chain and using those links to connect the remaining three chain pieces together into the required circular chain. This trick quickly gives us the successful solution.

Calculating Time Trick

The challenge is to calculate the current time given the following hints. To determine the current time, you need to determine one-quarter of the time from midnight to the present, and then add it to half the time from the present to the coming midnight.

At first view, this may be somewhat confusing. However, simple algebra will come to the rescue. If we let t represent the current time, we would be able to represent one-quarter of the time from midnight to the present as $\frac{t}{4}$. We can represent one-half the time from the present to midnight as $\frac{(24-t)}{2}$. This enables us to set up the following relationship: $\frac{t}{4}+\frac{(24-t)}{2}=t$, which reduces to $5t = 48$, so that $t = 9.6$ hours. Therefore, the current time is 9:36 AM.

Power Numbers

It is said that there is only one number between 2 and 200 trillion which is a perfect square, a perfect cube, and a perfect fifth power. The challenge is to find which number has these properties.

The number we seek has to be a number of 30th power, since $2 \times 3 \times 5 = 30$. We will choose a number and take it to the 30th power. If we choose $3^{30} = 205{,}891{,}132{,}094{,}649$, which is greater than 200 trillion. We then can try to see what $2^{30} = 1{,}073{,}741{,}824$, which is less than 200 trillion and therefore satisfies our challenge.

The Trick of Prime Ages

David tells his friend, who knows his age, that he has three cousins whose ages are prime numbers. The sum of his cousins' and his ages is 50 years. What are the ages of David and his three cousins?

When his friend indicates that David's age is 43, we immediately can see that the cousins are aged 2, 2, and 3, since the sum of $43 + 2 + 2 + 3 = 50$. Without knowing David's age, he would not be able to answer the question.

Guess the Weather

Your friend tells you that the weather has fluctuated over the past five days and was different each day. Each day at the same time, your friend measured the temperature. The product of the temperatures is 12°. What was the temperature on the five days?

The trick is to realize the temperatures could be positive or negative. One possibility is that the temperatures for the 5 days were 1°, –1°, 2°, –2°, and 3°, whose product is 12°.

Move the Coins According to the Rule

You might want to simulate this challenge in real life. In Figure 3-62, we show the locations of two pennies and two nickels in square units. The object of this challenge is to move the pennies to where the nickels are and move the nickels to where the pennies are. The only acceptable moves are that one can slide the coins to an empty square and one can jump over a coin if there is an empty square available on the other side. Be warned that this is not a simple task.

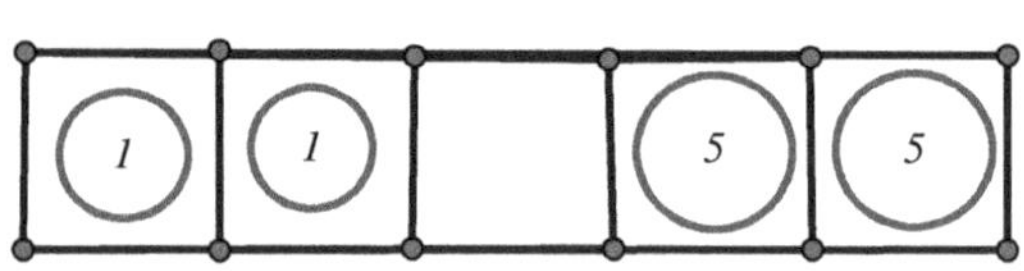

Figure 3-62

One strategy that can solve the problem is as follows:

1. Slide the penny to the empty square.
2. Jump the penny with the nickel.
3. Slide the other nickel to the empty square.
4. Jump the nickel with the penny.
5. Jump the other nickel with the other penny.
6. Slide the nickel to the empty square.
7. Jump the penny with the nickel.
8. Slide the penny which was just jumped over to the empty square.

You should now have the pennies where the nickels were and the nickels where the pennies were.

A Trick Fallacy

Three men enter a hotel and rent a suite of three rooms for \$300. After they are taken to their rooms, the manager of the hotel discovers he overcharged the men. The suite rents for only \$250. He then sends a bellhop upstairs with the \$50 refund. However, the dishonest bellhop decides to keep \$20 and returns only \$30 to the men. Now the rooms originally cost \$300, but the men had gotten \$30 returned to them. This means that they only paid \$270 for the rooms. The bellhop kept \$20. Therefore \$270 + \$20 = \$290. What happened to the extra \$10?

The surprising result is that there really is no missing \$10. The computation may be done in one of two ways: $(300–30) - 20 = 25$, or $250 + 20 + 30 = 300$. However the problem uses a faulty computation, that is, $(300–30) + 20 = 290$.

The Length of the Fish Puzzle

Consider a fish that has a tail as long as its head plus $\frac{1}{4}$ of the length of its body. Its body is $\frac{3}{4}$ of its total length. The head is 4 inches long. The challenge is to find the length of the fish.

To solve this problem, we let H represent the length of the head, let B represent the length of the body, let T the length of the tail, and let L the total length of the fish. From the information given, we know that $T = H + \frac{1}{4}B$, $B = \frac{3}{4}L$, and $H = 4$ inches. Furthermore, we know that L = H + B + T. Substituting in the last equation, we get

$$L = 4'' + \frac{3}{4}L + \left(H + \frac{1}{4}B\right) = 4'' + \frac{3}{4}L + 4'' + \frac{3}{16}L = 8'' + \frac{15}{16}L.$$

Therefore, $\frac{1}{16}L = 8''$, and $L = 128''$. So, we can conclude that the fish is 128 inches long.

The Problem of Party Participants

At an unusual party, every two guests shared a portion of French fries together, every three guests shared a portion of rice together, and every four guests shared a quantity of meat together. If there were 65 dishes altogether at the party, how many guests were in attendance?

An algebraic solution is helpful here. We let x equal the number of guests. We can represent the number of dishes as follows:

$$\frac{x}{2} + \frac{x}{3} + \frac{x}{4} = 65,$$

which gives us $6x + 4x + 3x = 65 \times 12$, and thus, $x = 60$, which is the number of guests at the party.

A Board Covering Challenge

The board shown in Figure 3-63 has 49 squares, with one marked with an X. The task is to cover the remaining 48 squares with 24 two-square tiles, as shown in the figure.

Careful inspection of the board shows that the 48 squares to be covered include 23 unshaded squares and 25 shaded squares. The tile used to cover two squares will always cover a shaded and an unshaded square. That procedure will leave two uncovered squares that cannot be filled as they are not of two different shades. Therefore, no solution is possible. This is a fine example of how logical reasoning helps respond to the challenge.

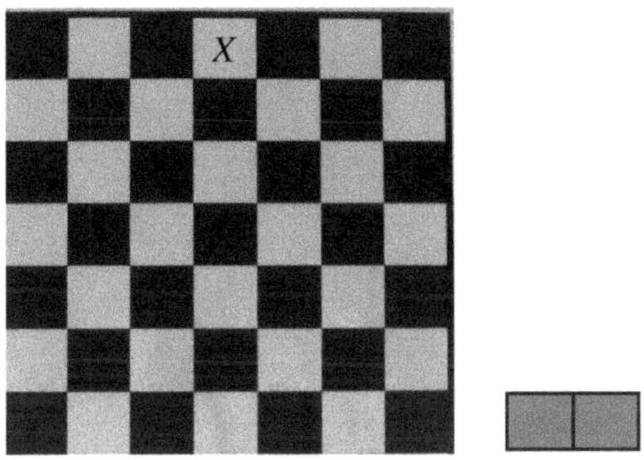

Figure 3-63

Identifying a Quarter Region of an Unusual Polygon

Figure 3-64 shows a shape which is $\frac{3}{4}$ of a square. The challenge is to separate this polygon into quarters.

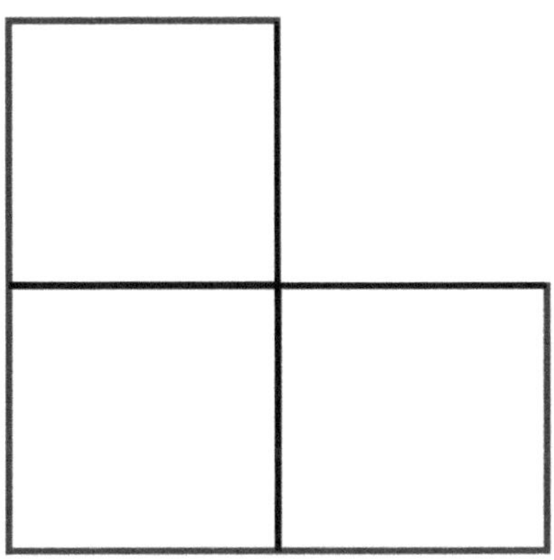

Figure 3-64

One method for denoting four quarter regions in this strange polygon is to partition each of the three given large squares into four quarters. Then we have twelve smaller squares, where three of these smaller squares make $\frac{1}{4}$ of the original polygon. Figure 3-65 shows four quarter regions, each comprised of three smaller squares.

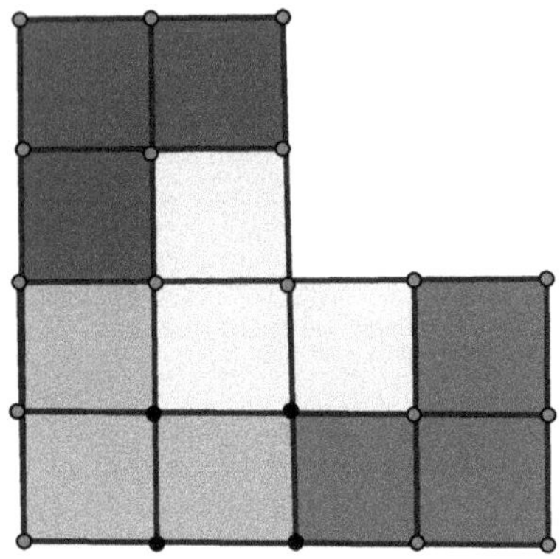

Figure 3-65

An alternate solution is shown in Figure 3-66, where the polygon has been partitioned into twelve regions, and three of the triangular regions represent $\frac{1}{4}$ of the original polygonal figure.

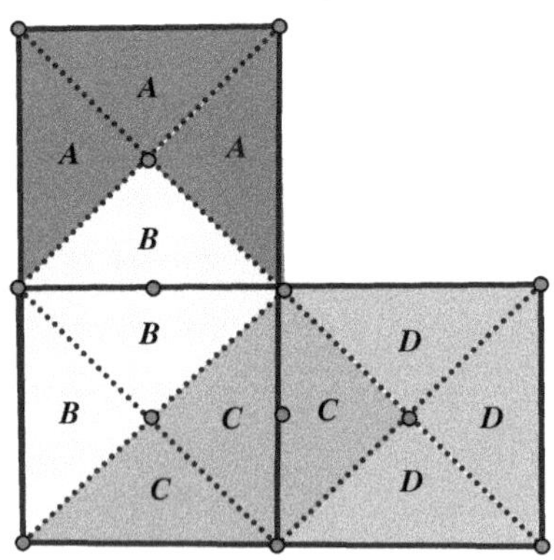

Figure 3-66

Generating Squares Creatively

Figure 3-67shows three squares. The challenge is to remove three of the line segments and produce a configuration that has five squares.

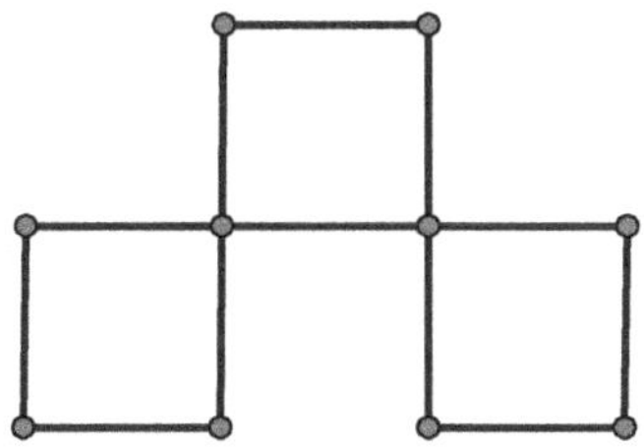

Figure 3-67

In Figure 3-68, we have removed three line segments from their former positions (represented by dots) and placed them where indicated by the bold marks. We now have four small squares and the large square encompassing the four squares, which results in five squares.

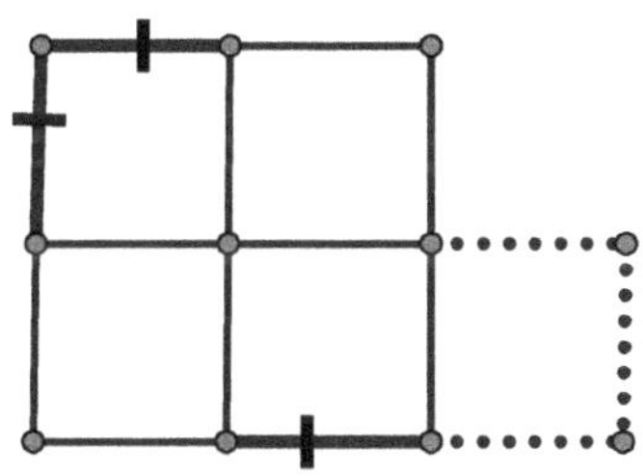

Figure 3-68

We use logical reasoning throughout mathematics. This chapter exhibited the some of the aspects of logic as seen with number relationships, arithmetical manipulations, geometrical challenges, and general thinking.

Chapter 4

Geometric Wonders

Geometry represents the most visual aspect of elementary mathematics. It also exhibits a lot of unusual relationships. Many of these relationships can be justified with cleverness and an element of sophistication. There are also aspects of visual mathematics where we use proofs to justify the relationship. This chapter presents unusual geometric relationships; where it is appropriate, a proof is provided. A special feature of this chapter is that the surprising relationships are not usually encountered in the study of geometry. However, a key to appreciating this chapter is to recognize the beauty of the relationships presented.

Surprising Oddity: A Rope Around the Earth

We present here and unanticipated oddity about a rope tied around the Earth. Consider the globe of the Earth with a rope wrapped tightly around the equator, as shown in Figure 4-1. The rope will be about 24,900 miles long. The rope is lengthened by exactly 1 yard and then positioned around the equator so that it is uniformly spaced above the globe. The question is, will a mouse fit under the rope?

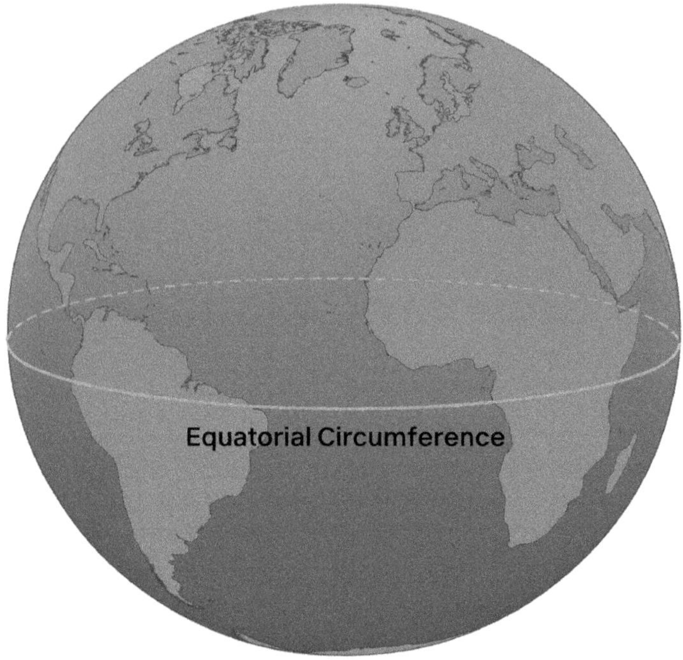

Figure 4-1

The traditional way to determine the difference between the circumferences is to find the difference between the radii. Let R be the length of the radius of the circle formed by the rope (circumference $C + 1$) and let r the length of the radius of the circle formed by the earth (circumference C), as shown in Figure 4-2.

The familiar circumference formulas give us:

$$C = 2\pi r, \text{ or } r = \frac{C}{2\pi} \quad \text{and} \quad C+1 = 2\pi R, \text{ or } R = \frac{C+1}{2\pi}.$$

We need to find the difference of the radii:

$$R - r = \frac{C+1}{2\pi} - \frac{c}{2\pi} = \frac{1}{2\pi} \approx .159 \text{ yards} \approx 5.7 \text{ inches.}$$

Wow! There is a space of **over 5½ inches** for a mouse to crawl under. You must really appreciate this astonishing result.

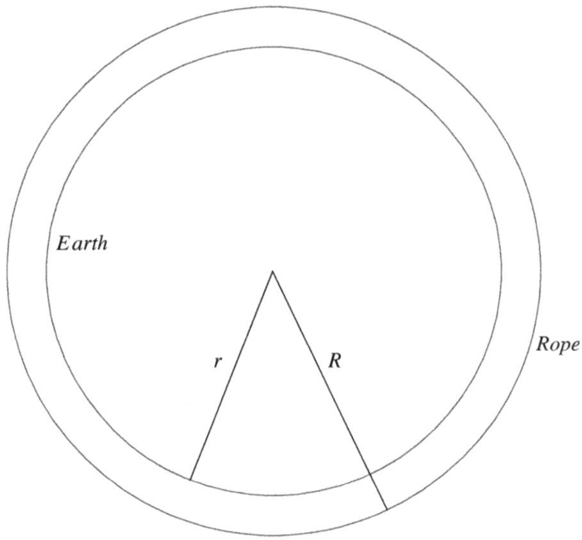

Figure 4-2

Imagine, lengthening the 24,900-mile rope by 1 yard lifted the rope off the equator about 5½ inches!

Consider the original problem shown in Figure 4-2. You should realize that the solution was independent of the circumference of the earth, since the end result did not include the circumference in the calculation. It only required calculating $\frac{1}{2\pi}$.

Here is a really nifty solution using an extreme case. Suppose the inner circle (that is, the Earth) is very small, so small that it has a zero-length radius (that means it is actually just a point). We were required to find the difference between the radii, which, when $r = 0$, is $R - r = R - 0 = R$.

So, all we need to find is the length of the radius of the larger circle, and our problem will be solved. With the circumference of the smaller circle now 0, we apply the formula for the circumference of the larger circle:

$$C+1=0+1=2\pi R, \text{ then } R=\frac{1}{2\pi}.$$

An Unusual Technique to Show the Angle Sum of a Triangle

It is widely accepted that for a plane triangle the sum of the angles is 180°. This is the basis for Euclidean geometry, and ought to be understood by all. Most people know that when they make one complete revolution, that represents 360°. There is nothing sacrosanct about this measure, other than that it is generally accepted and so used. That is, we agree that a complete revolution is 360°. So, a half-revolution is 180°. How does the angle sum of a triangle relate to this? The simplest and perhaps the most convincing way to demonstrate this angle sum is to tear the three vertices from a paper triangle and place them together to form a straight line. The straight line represents one-half of a complete revolution, hence, the 180°.

It is perhaps neater to fold the triangle. Cut a conveniently large scalene triangle from a piece of paper. Then fold one vertex so that it touches the opposite side and so that the crease is parallel to that side, as shown in Figure 4-3.

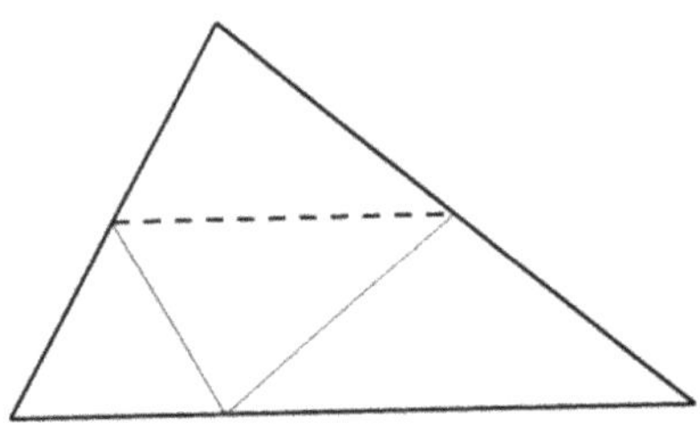

Figure 4-3

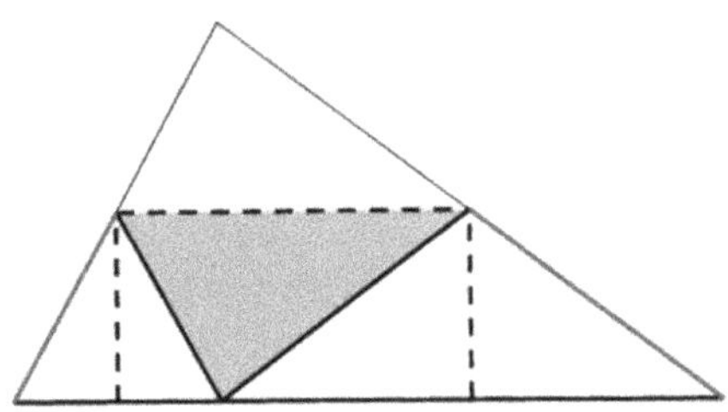

Figure 4-4

Then fold the remaining two vertices to meet the first vertex at a common point, as shown in Figure 4-4. You will notice that the three angles of the triangle come together and form a straight line, and hence have an angle sum of 180°, as we show in Figure 4-5.

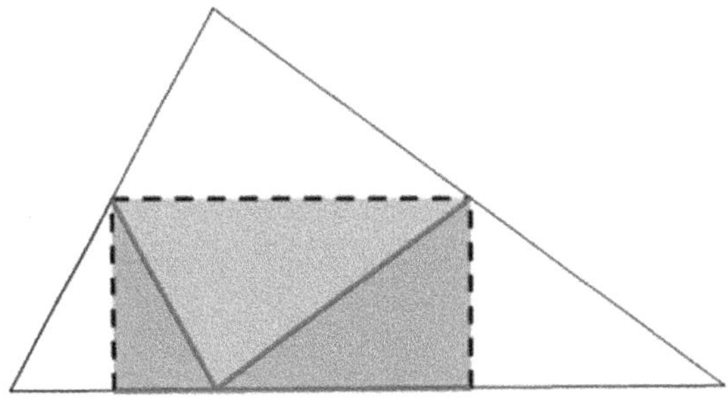

Figure 4-5

However, it is also nice to show why this folding procedure has the vertices meet at a point on the side of the triangle. Establishing this phenomenon is tantamount to proving the theorem establishing the angle sum of a triangle.

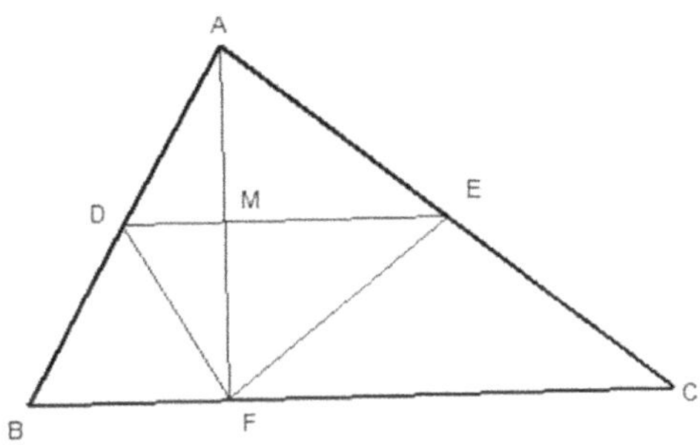

Figure 4-6

The proof of this theorem for the interested reader follows directly from the paper-folding exercise. We fold the top vertex along a parallel crease so that $DE \parallel BC$ and $AF \perp ED$ at M, as shown in Figure 4-6. Since $MF = AM$ (that is, M is the midpoint of AF), D and E are midpoints of AB and AC, respectively, since a line parallel to one side of a triangle (either ΔBAF or ΔCAF) and bisecting a second side (AF) of the triangle also bisects the third side. It is then easy to show that since $AD = DF$, $BD = DF$, and similarly, $EF = EC$, so that the folding over of vertices B and C would fit at F, forming a straight line along BCF.

A Simple Solution to a Curious Problem

There are problems that appear very simple but whose solutions are extraordinary. This makes for a mind-opening experience. We have four congruent squares shown in Figure 4-7, and lines AD and CE intersect at point B at an angle of 45°. The question is, how do we know that this angle is actually 45°? This is a rather strange approach because usually the task is to find the measure of angle ABC. Here, we just need to know how to justify 45°.

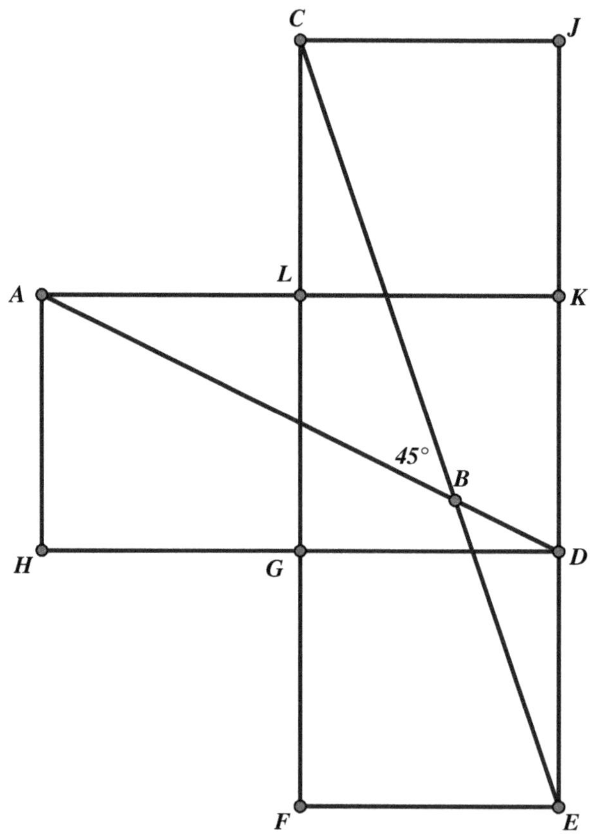

Figure 4-7

There are many procedures that can be used to determine the measure of angle *ABC*. We will simply draw lines *CH* and *EH*, as shown in Figure 4-8, and find that they are perpendicular for the following reason. We can prove $\Delta HGC \cong \Delta EDH$ so that $\angle HCG = \angle EHD$. We also have $\Delta NCL \cong \Delta AHN$, and so $\angle HCG = \angle AHN$. Therefore, $\angle AHN = \angle EHD$. Now, since $\angle AHN + \angle CHD = 90°$, by substitution we also get $\angle EHD + \angle CHD = 90°$. Therefore, we have right triangle *EHC* with $CH = EH$, and thus, $\angle CEH = 45°$. Clearly $AD \| EH$, giving us the measure of $\angle ABC = 45°$.

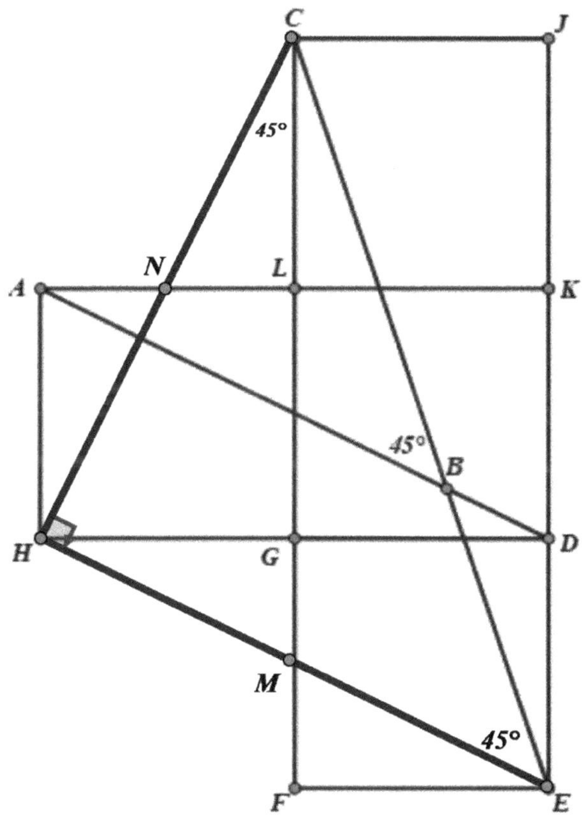

Figure 4-8

Unusual Comparison of Areas of Squares

We can show in a clever way that the area of square *JKLN* has $\frac{1}{5}$ the area of square *ABCD*, as can be seen in Figure 4-9, where *E*, *F*, *G*, and *H* are the midpoints of the sides of the square *ABCD*.

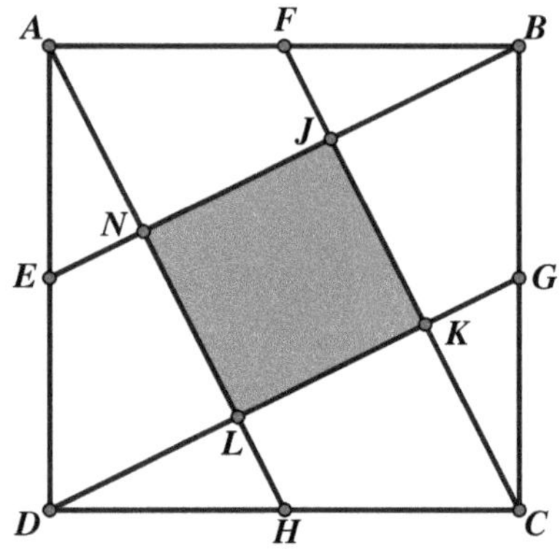

Figure 4-9

The cleverness comes from the fact that we would like to create a situation which demonstrates that square *JKLN* is $\frac{1}{5}$ of a constructed area. To do that, we will extend the sides of square *JKLN* so that we have the situation shown in Figure 4-10, where we then make $\angle P$, $\angle Q$, $\angle R$, and $\angle M$ right angles.

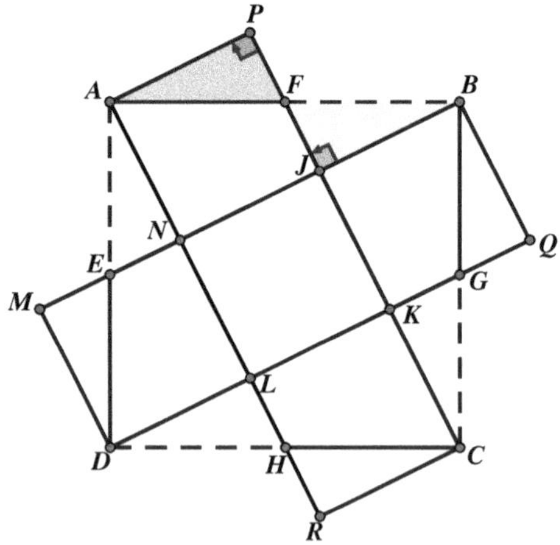

Figure 4-10

Just as we can show that $\Delta FBJ \cong \Delta FAP$, we also have $\Delta NEA \cong \Delta EMD$, $\Delta LHD \cong \Delta RHC$, and $\Delta KGC \cong \Delta QGB$. Therefore, the area of the cross is equal to the area of the original square $ABCD$. Since square $JKLN$ is $\frac{1}{5}$ of the area of the cross, we can conclude that square $JKLN$ has $\frac{1}{5}$ of the area of square $ABCD$.

The Constant Chord

In Figure 4-11, circles O and R intersect at points E and F. Point P is on circle R, and PF and PE intersect circle O at points A and B. We then choose any other point Q on circle R and draw the secants from Q through points E and F to intersect circle O at points D and C, respectively. Regardless of where Q is positioned on circle R, the chord CD will always be the same length as the chord AB. This is truly an astonishing oddity!

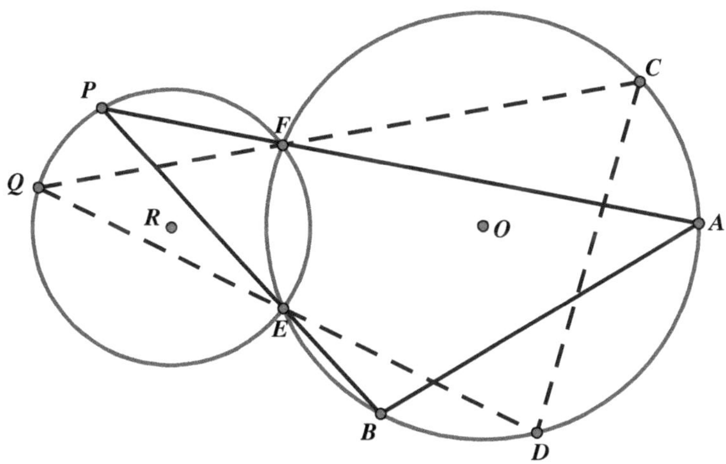

Figure 4-11

To demonstrate why this unexpected equality occurs, we begin by noting that the pair of vertical angles are equal, that is, $\angle PFQ = \angle AFC$ and $\angle PEQ = \angle BED$. However, $\angle PFQ = \angle PEQ$, since they are both measured by $\overset{\frown}{EF}$. Therefore, since $\angle AFC = \angle BED$, we know that $\overset{\frown}{AC} = \overset{\frown}{BD}$. Thus, by adding $\overset{\frown}{AD}$ to each of these arcs, we get $\overset{\frown}{AB} = \overset{\frown}{CD}$, and it follows that $AB = CD$, which is what we set out to show.

Two Unusual Equal Chords in a Circle

In Figure 4-12, circle O passes through the center of circle P, and the circles intersect at points A and B. Curiously, the tangent to circle O at point A intersects circle P at point C and creates a segment such that $AC = AB$.

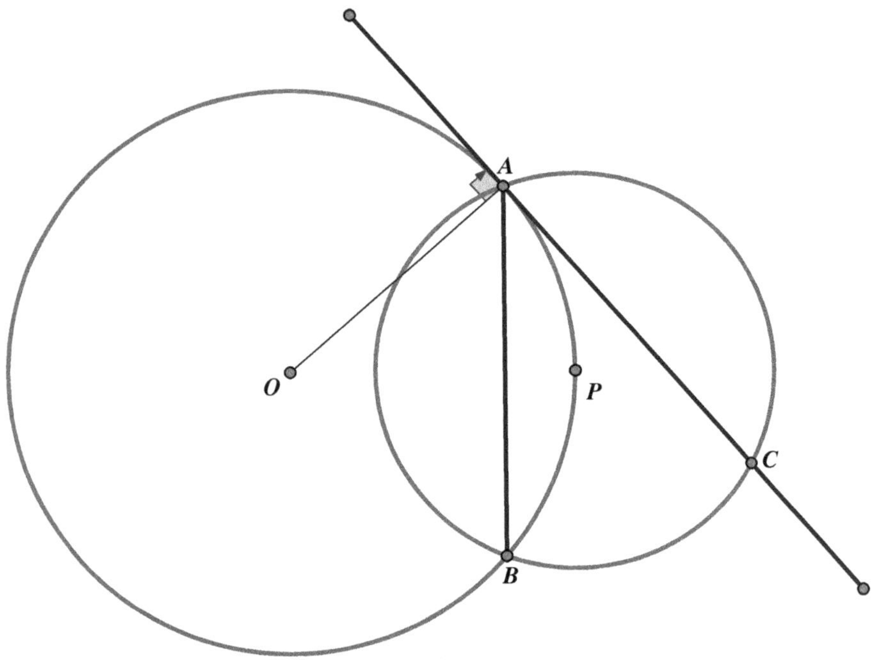

Figure 4-12

This interesting coincidence can be easily justified in the following way. We draw radii *AP*, *BP*, and *CP*, as shown in Figure 4-13. Since the radii $AP = BP$, we know that the arcs $\overset{\frown}{AP} = \overset{\frown}{BP}$, which gives us $\angle PBA = \angle PAB$. Furthermore, $\angle PBA = \angle PAC$, since both angles are measured by arc *AP*. Thus, the isosceles triangles $\Delta APB \cong \Delta APC$. Hence, we have $AC = AB$.

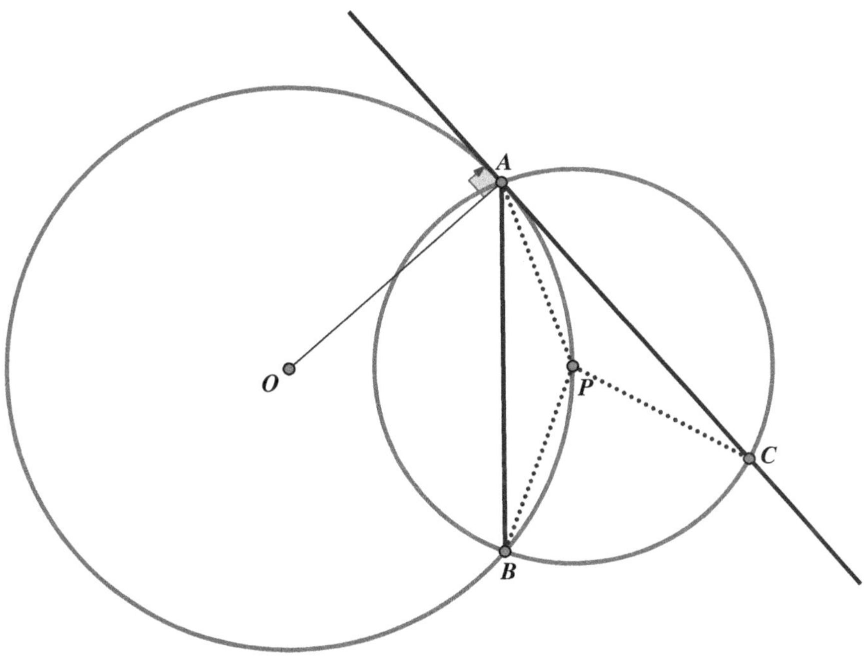

Figure 4-13

A Technique to Find the Sum of the Angles of a Pentagram

There are various ways to establish that the sum of the angles of a pentagram is 180°, including complicated proofs based on the pentagon within the pentagram. However, here is a clever way of determining the angle sum of the pentagram. Consider the pentagram *ABCDE*, shown in Figure 4-14, where the angle measures are indicated numerically.

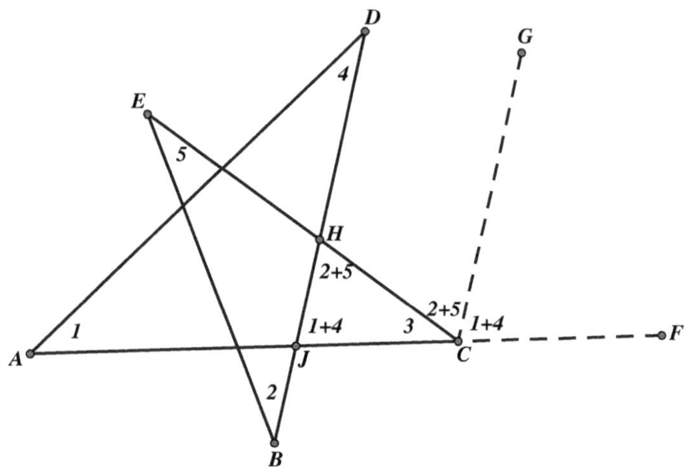

Figure 4-14

Begin by noting that an exterior angle of the triangle is equal to the sum of the two remote interior angles. For ΔADJ we have $\angle DJC = 1 + 4$, and for ΔHEB we have $\angle BHC = 2 + 5$. After extending *AC* to point *F*, we construct *GC* parallel to *BD*, whereupon $\angle GCH = \angle CHJ = 2 + 5$, and $\angle GCF = \angle CJH = 1 + 4$. We then have $\angle ACE + \angle GCE + \angle FCG = 180°$, represented by $1 + 2 + 3 + 4 + 5 = 180°$, which is the sum of the angles of the pentagram.

An Unusual Trick Geometric Solution

Here is a very unusual solution to a problem that at first glance seems overwhelming. Consider the square *ACDE*, shown in Figure 4-15, with point *B* in the interior so that $CB = 1$, $AB = 2$, and $EB = 3$. The challenge is to find the measure of angle *ABC*.

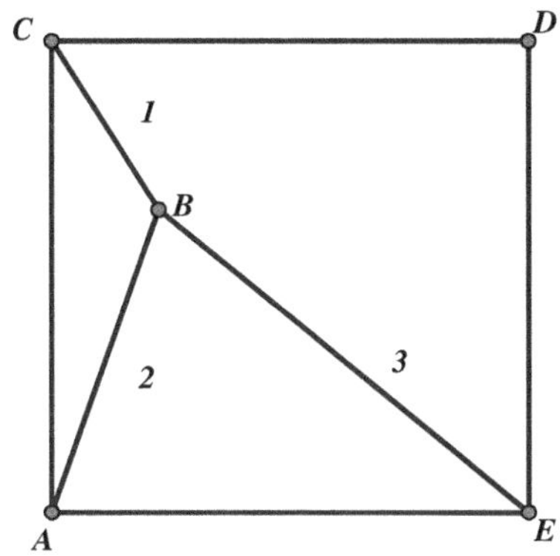

Figure 4-15

The first step in the solution is unusual. We rotate square *ACDE* around point *A* in a counterclockwise direction by 90°. In Figure 4-16, we show the gradual rotation, which results in a rectangle *DD′C′E*, as can be seen in Figure 4-17. We then draw line segment *B′B*.

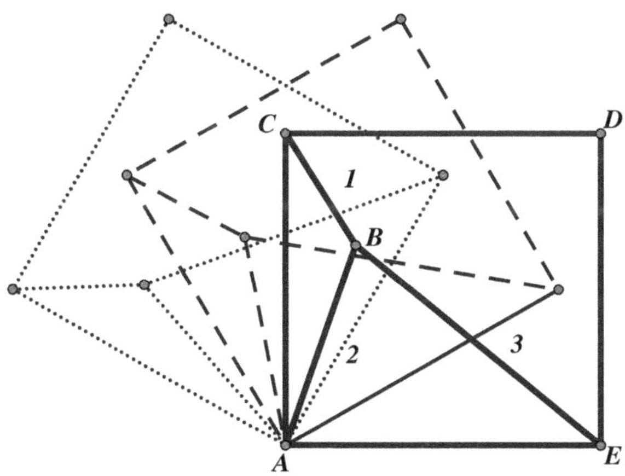

Figure 4-16

With the rotation of 90°, we find that $\angle BAB' = 90°$. Furthermore, right triangle *BAB′* is isosceles so that $\angle ABB' = 45°$. By the Pythagorean theorem, we find that $BB' = 2\sqrt{2}$. From the converse of the Pythagorean theorem applied to triangle *BCB′*, since $1^2 + 3^2 = \left(2\sqrt{2}\right)^2$ we have $\angle CBB' = 90°$. Therefore, $\angle ABC = \angle CBB' + \angle ABB' = 90° + 45° = 135°$, which fulfills our initial task.

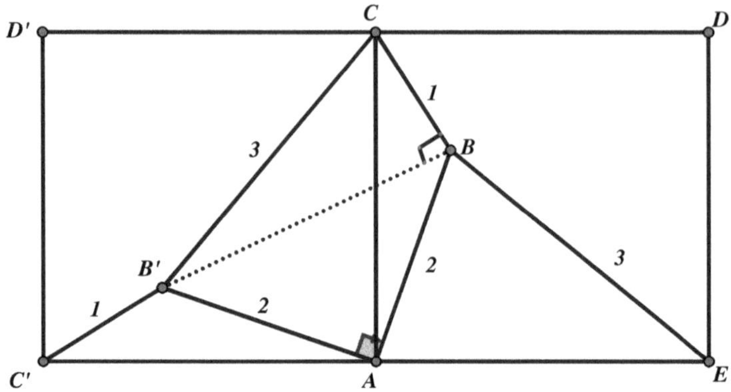

Figure 4-17

Creating Similar Triangles

Here is a curious challenge. Given a rectangle, use two lines to create three similar triangles.

There are two possibilities to consider. One is good for any rectangle, such as the one presented in Figure 4-18, where we show three right triangles $\Delta ABE \sim \Delta BCE \sim \Delta DAC$ that have equal angles $\angle ABE = \angle ACB = \angle DAC$, since they are each complementary to $\angle BAC$.

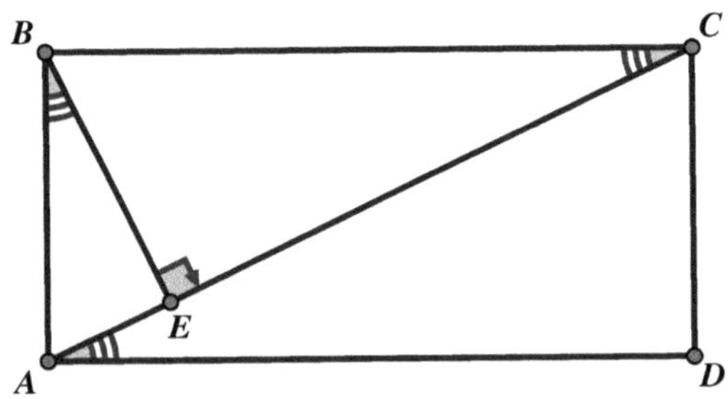

Figure 4-18

Another possible way to create three similar triangles would require a rectangle where the length is more than twice the width, such as the rectangle shown in Figure 4-19, where we select a point E on side BC such that $\angle AED = 90°$. Since $\angle BEA$ and $\angle CDE$ are both complementary to $\angle DEC$, they are equal. From the parallel lines AD

and *BC*, we have the alternate-interior angles $\angle AEB = \angle DAE$, and so the right triangles $\Delta AEB \sim \Delta EDC \sim \Delta ADE$.

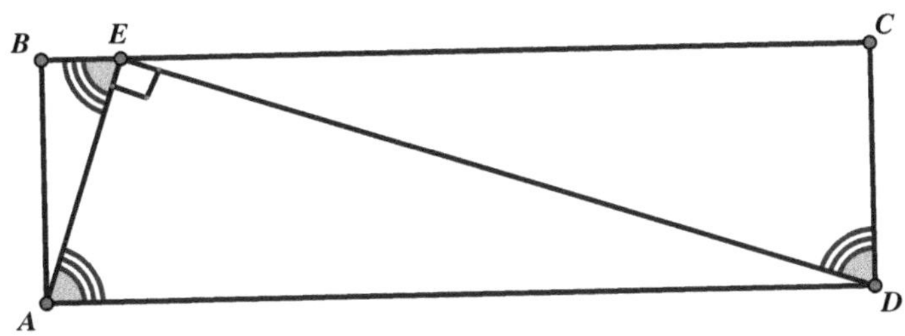

Figure 4-19

Creating Ten Non-Overlapping Triangles with Two Lines and a Pentagram

In Figure 4-20, we have a randomly drawn pentagram *ABCDE*. We would like to draw two straight lines that would partition this pentagram, which currently shows five non-overlapping triangles, and create five more non-overlapping triangles.

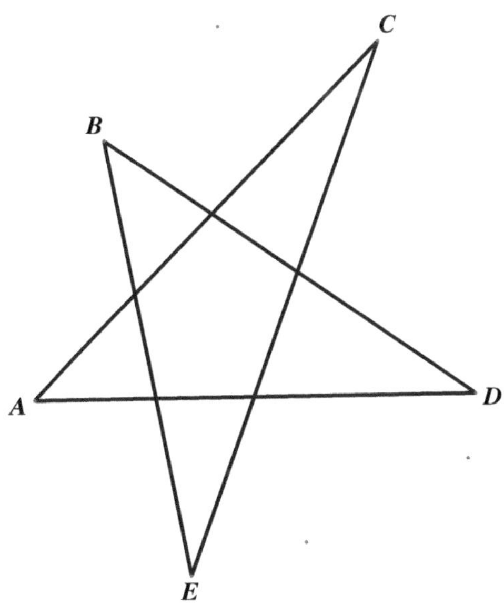

Figure 4-20

The two lines *FG* and *FH*, shown in Figure 4-21, create a second pentagram, namely, *NIPFC.*

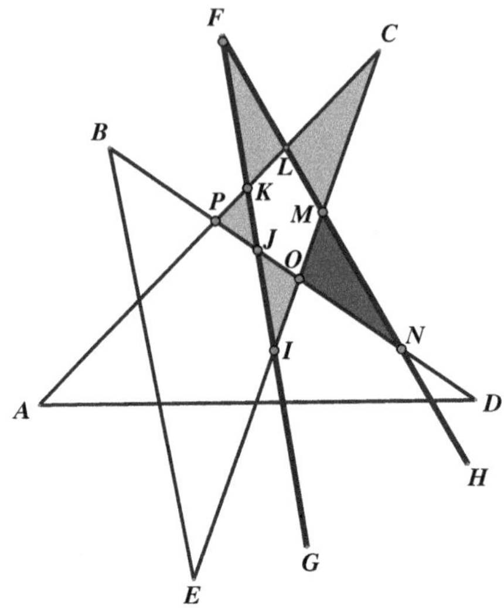

Figure 4-21

When we count the non-overlapping triangles in Figure 4-22, we find that there are exactly ten such triangles, which was the initial challenge.

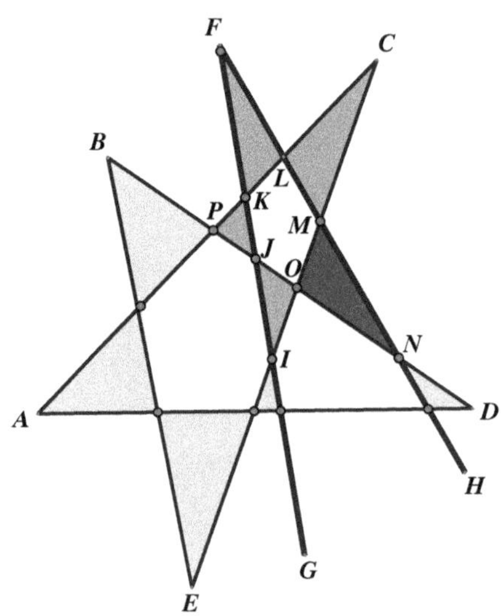

Figure 4-22

Creating Seven Non-Overlapping Triangles with Three Lines and a Line Intersecting Two Parallel Lines

In Figure 4-23, we have a line intersecting two parallel lines, and we seek to create seven non-overlapping triangles by intersecting the three given lines with three other lines.

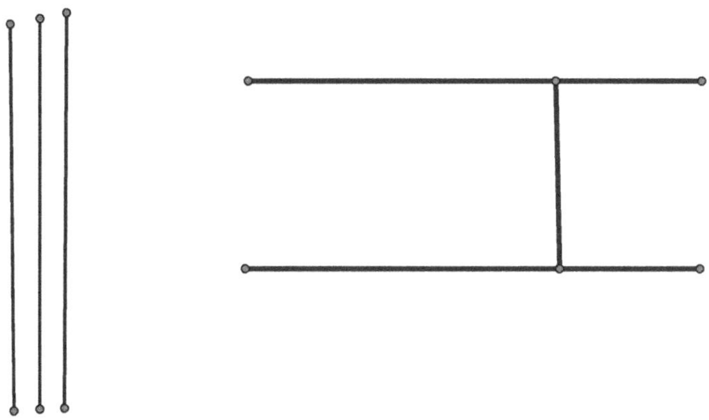

Figure 4-23

Once again, we will create a pentagram that will lead to the desired solution, as we can see in Figure 4-24.

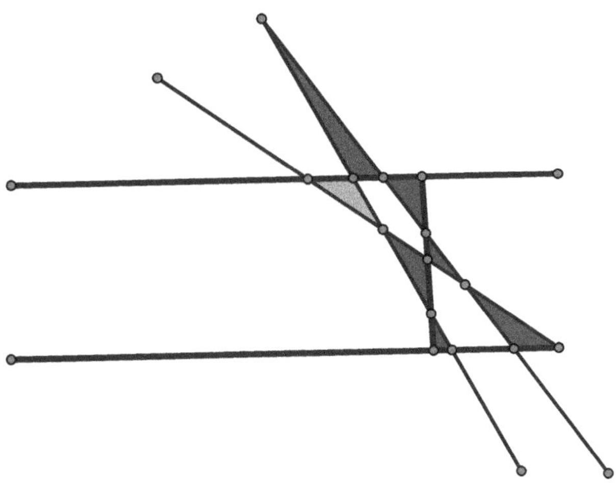

Figure 4-24

Creating Nine Non-Overlapping Triangles with Three Lines Crossing Four Lines in a W Shape

In Figure 4-25, we seek to determine how the three lines shown can cross the W shape and create nine non-overlapping triangles.

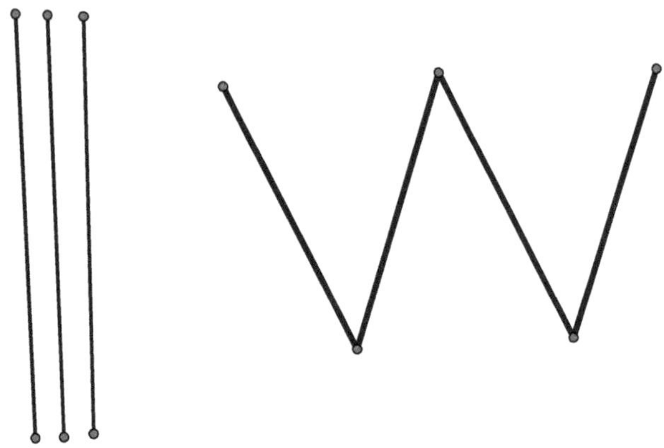

Figure 4-25

Once again, we strategically cross the given four lines with the three movable lines to create nine non-overlapping triangles, as seen in Figure 4-26.

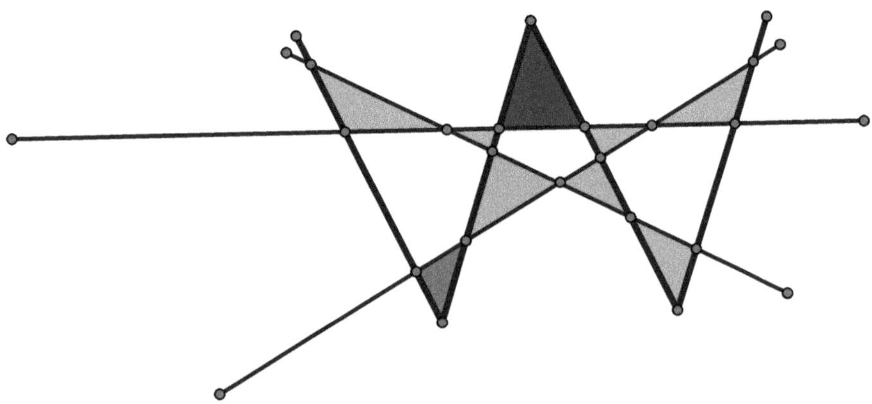

Figure 4-26

Identifying Two Squares

In Figure 4-27, a square is formed and partitioned into four quarters using 12 sticks. The challenge is to remove two sticks and be left with exactly two squares. Try to do this without looking ahead at the answer.

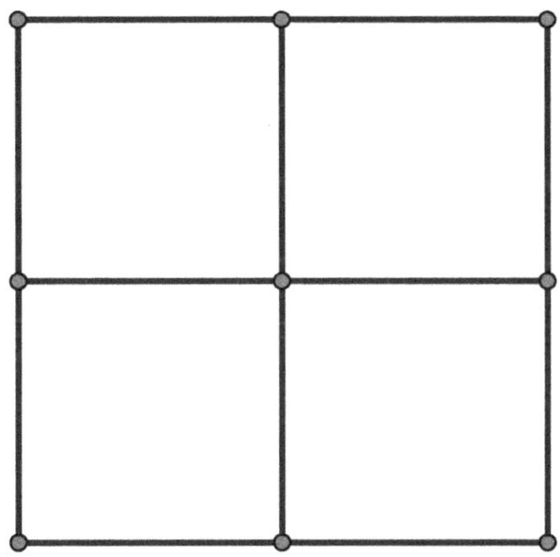

Figure 4-27

We show the solution to the question in Figure 4-28, where two sticks have been removed and only two squares remain.

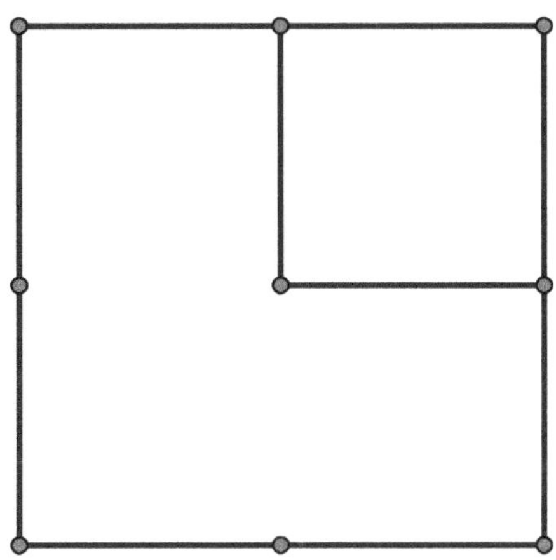

Figure 4-28

Such strategies can come in handy, as our mental capacity is guided by past experience.

From Five Squares Create Four Squares

Given the arrangement of five squares as shown in Figure 4-29, remove two sides from the five-square arrangement to leave only four squares.

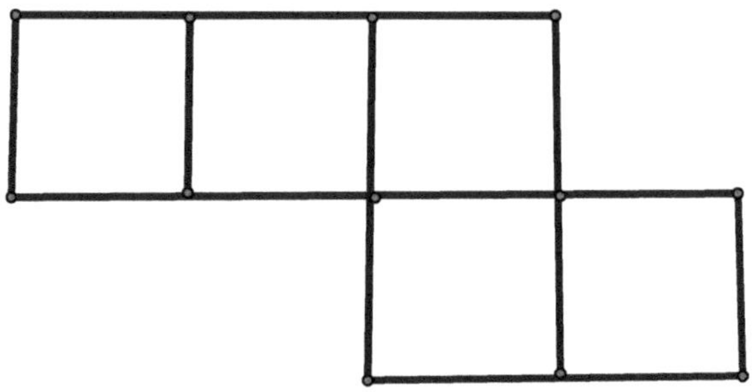

Figure 4-29

Figure 4-30 exhibits one way that this can be done, where two squares were eliminated and the two sides created a new square. Notice, there are only four squares now.

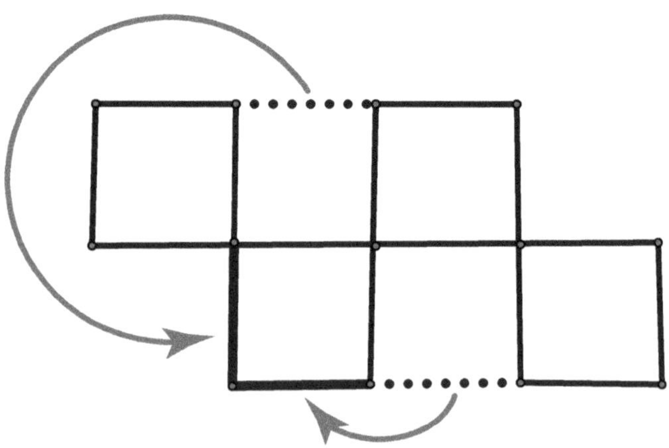

Figure 4-30

Cutting Two Squares to Create a Larger Square of Equal Area

Figure 4-31 shows two squares, *ABCD* and *AEFG*. The task is to cut these two squares into five or fewer pieces to create another square equal in area to the two squares.

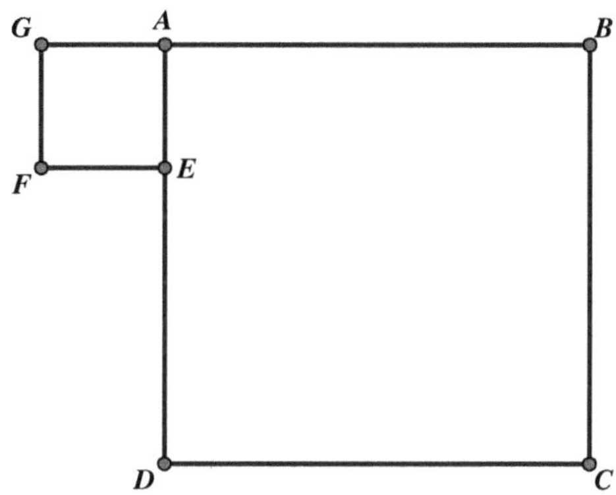

Figure 4-31

Begin by marking off *H* on *AB* such that $HB = GF$ and then drawing *HF* and *HC*, as shown in Figure 4-32. Since $GA + AH = AH + HB = AB = BC$, we have $\Delta FGH \cong \Delta HBC$. Also, $\angle FHG + \angle CHB = 90°$ so that

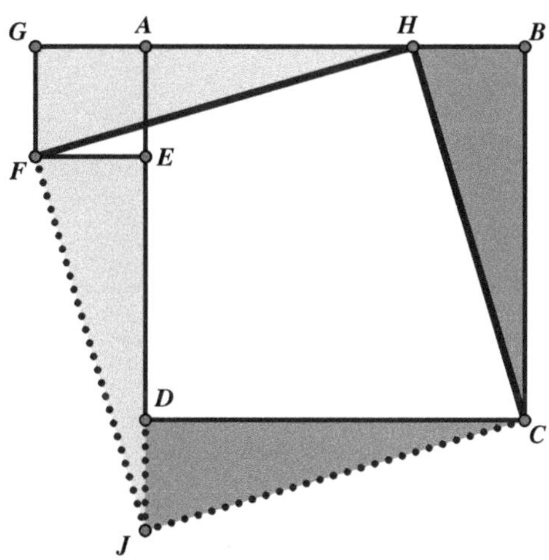

Figure 4-32

$\angle CHF = 180° - \angle FHG + \angle CHB = 90°$. So, we have two sides of the future square, namely *HF* and *HC*, and a right-angle *CHF*. We then rotate ΔHBC counterclockwise 90° at point *C* to get ΔJDC. In a similar fashion, we can show that $\Delta FEJ \cong \Delta JDC$. Therefore, the square *ABCD* has the same area as square *FHCJ*. And so, we have shown that the two given squares could be cut into no more than five pieces to create one square with equal area to the sum of the two original squares.

Semicircle and Quarter Circle Define an Angle

In Figure 4-33, square *ABCD* has a quarter circle inscribed with center at vertex *C* and a semicircle inscribed with diameter *BC*. Point *P* is on semicircle *BC*, and *CP* intersects arc *BD* at point *R*. If $\angle PBR = 30°$, we need to find the measure of $\angle ABR$.

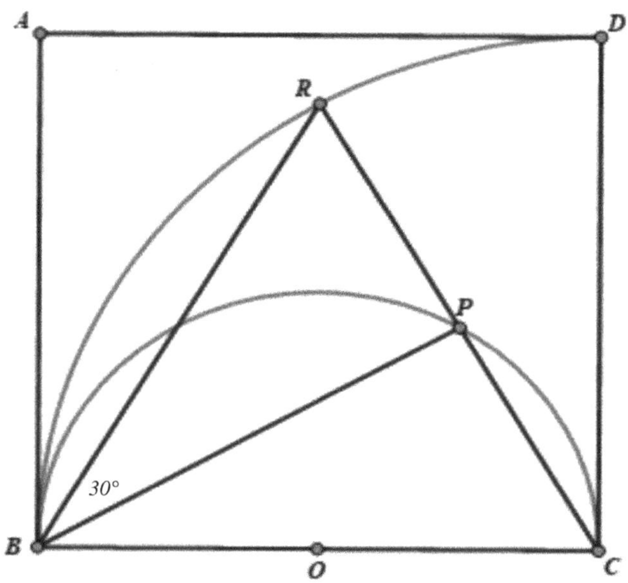

Figure 4-33

There are a variety of ways to show that $\angle PBR = \angle ABR$, which immediately solves the problem by indicating that $\angle ABR = 30°$.

Considering the angles at vertex B, we have $\angle PBR = \angle ABC - \angle ABR - \angle PBC$. However, $\angle ABR = \frac{1}{2}\angle BCR$, since both angles are measured by $\widehat{BR}$. Substituting that into the previous equation, we get $\angle PBR = 90° - \frac{1}{2}\angle BCR - \angle PBC$. Since triangle PBC is inscribed in a semicircle, it is a right triangle, and therefore, $\angle PBC = 90° - \angle BCR$. We substitute that into the previous equation and get $\angle PBR = 90° - \frac{1}{2}\angle BCR - (90° - \angle BCR) = \frac{1}{2}\angle BCR = \angle ABR$. Therefore, in this situation, we have $\angle PBC = \angle ABR = 30°$.

Extremes Solve the Problem

Here we have a classic example where considering an *extreme* trivializes the original challenge. In Figure 4-34, AB is a chord of the larger of the two concentric circles and is tangent to the smaller circle at point T. Find the area of the lighter shaded region between the two circles if $AB = 8$.

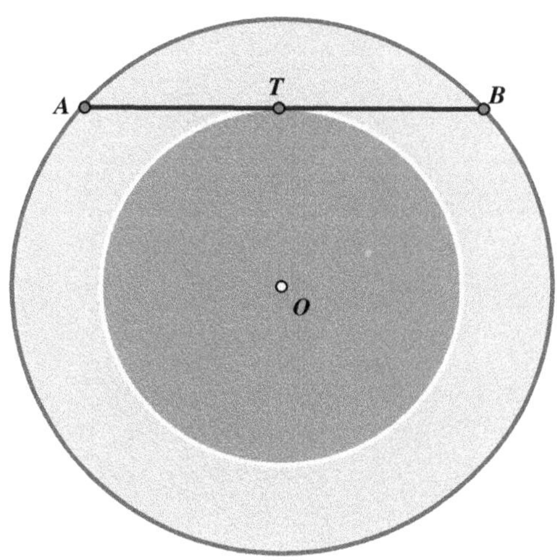

Figure 4-34

To appreciate the technique of using extremes, we will first consider the traditional method of solving this problem and then compare it to the extreme-based solution. We know that a radius is perpendicular to a tangent at the point of contact, T. Furthermore, a radius perpendicular to a chord divides the chord into two equal segments. Thus, as shown in Figure 4-35, we have $AT = BT = 4$. We also know that the area of the region between the two circles (the "doughnut" shape) can be found by obtaining the difference between the areas of the two circles. Thus, the area of this region between the two circles equals $\pi R^2 - \pi r^2 = \pi\,(R^2 - r^2)$. Now, since $OC = R$ and $OT = r$, $CT = (R - r)$ and $DT = (R + r)$.

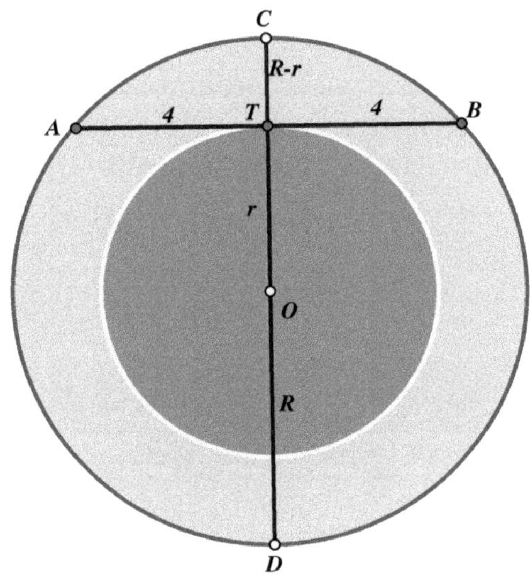

Figure 4-35

Since the product of the segments of two intersecting chords is equal, we have $(R - r)(R + r) = 4 \cdot 4$, so that $R^2 - r^2 = 16$. Therefore, the area of the region between the two circles equals $\pi R^2 - \pi r^2 - \pi\left(R^2 - r^2\right) = 16\pi$ square units.

We can also solve this problem by drawing a line segment OA to create a right triangle ATO, in which $(OA)^2 = (OT)^2 + 4^2$, and once again $R^2 - r^2 = 16$, so that the area of the region between the two circles equals $\pi R^2 - \pi r^2 - \pi\left(R^2 - r^2\right) = 16\pi$ square units.

However, we can also look at this problem by considering an extreme case. Let's assume that the smaller circle gets smaller and smaller until it becomes a point and coincides with point O. Then AB becomes the diameter of the larger circle. The area of the region between the two circles becomes the area of the larger circle, which equals $\pi R^2 = 16\pi$, which is a considerably more elegant solution to the problem. Considering an extreme provides us with the sought value, but it does not prove that this value is independent of the involved radii.

Extremes Again Solve the Problem

Considering extremes can be quite helpful when determining the sum of the exterior angles of a polygon, such as the one shown in Figure 4-36, where the exterior angles are marked u, v, w, x, y, and z. Rather than drawing numerous auxiliary lines, we can take the figure and gently reduce its size, as shown in the successive pictures of Figure 4-36, where the polygon eventually becomes so small that it is a point. The sum of the exterior angles surrounding the point is 360°. Throughout this process, the angles' measures did not change, only their initial positions.

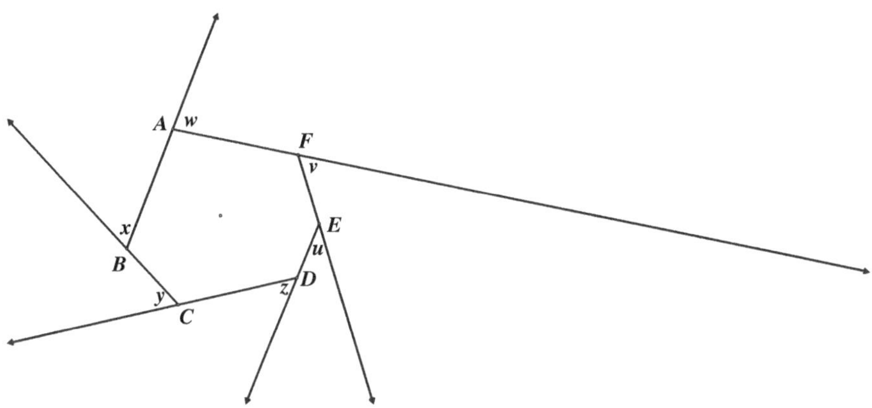

Figure 4-36 (*Continued*)

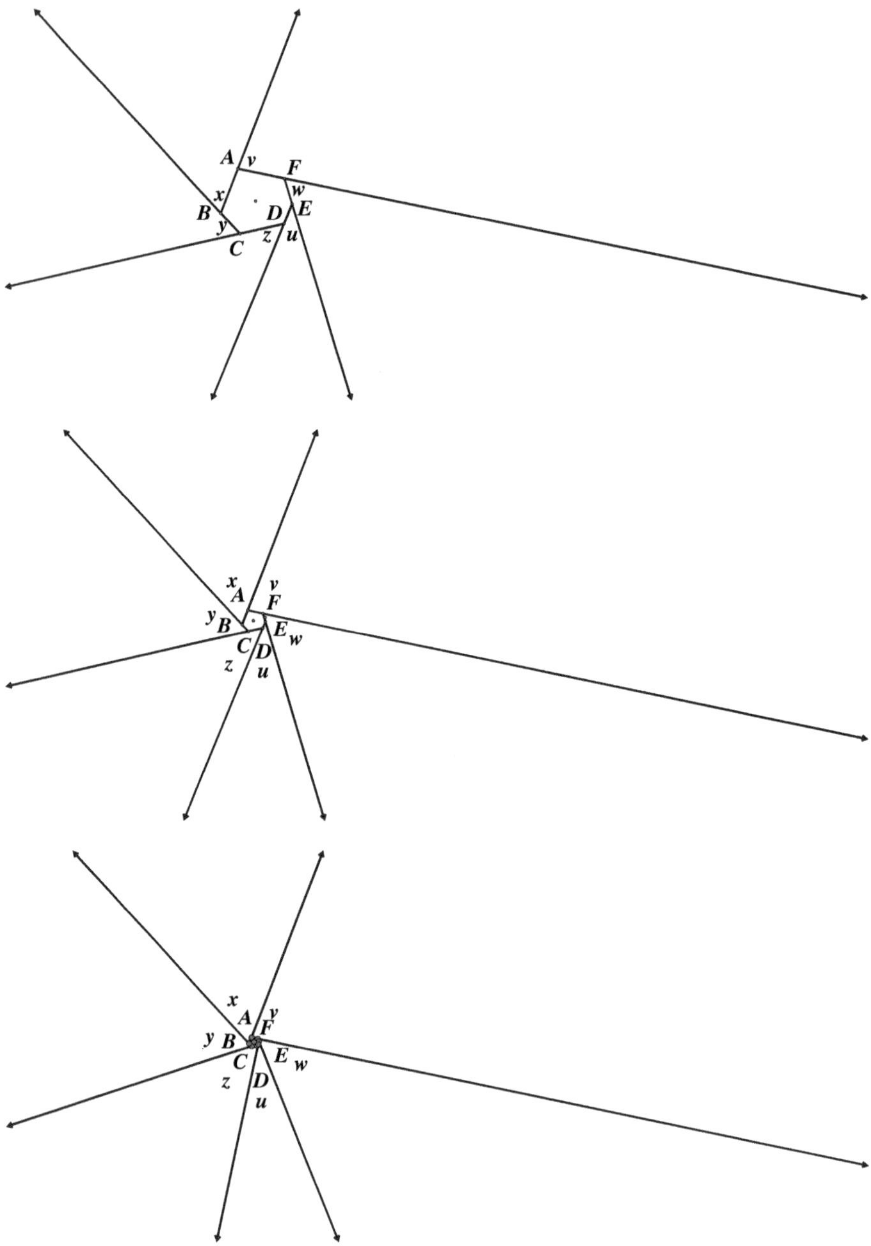

Figure 4-36

The Rectangle Puzzle

Figure 4-37 shows a rectangle partitioned into nine congruent rectangles. If the area of the large rectangle is 45, what is the perimeter of this large rectangle?

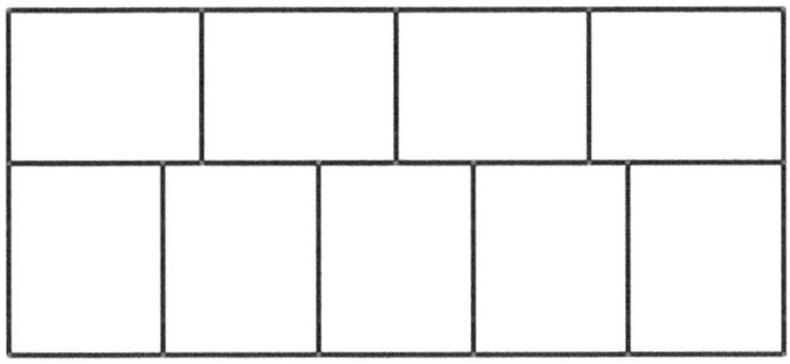

Figure 4-37

This problem deals with proportion. The width of five small congruent rectangles equals the length of four rectangles, that is, $5w = 4l$, and we have $\frac{l}{w} = \frac{5}{4}$. Furthermore, the area of each of the nine rectangles is $lw = \frac{45}{9} = 5$. Since $l = \frac{5w}{4}$, we have $\left(\frac{5w}{4}\right)w = 5$, and then $w = 2$, therefore, $l = \frac{5}{2}$. We are now able to calculate the perimeter of the large rectangle:

$$2(5w + w + l) = 2\left[5(2) + 2 + \frac{5}{2}\right] = 29.$$

A Mysterious Shaded Region

In Figure 4-38, the sum of the two diameters of the smaller semicircles is equal to the diameter of the larger semicircle. If the ratio of the

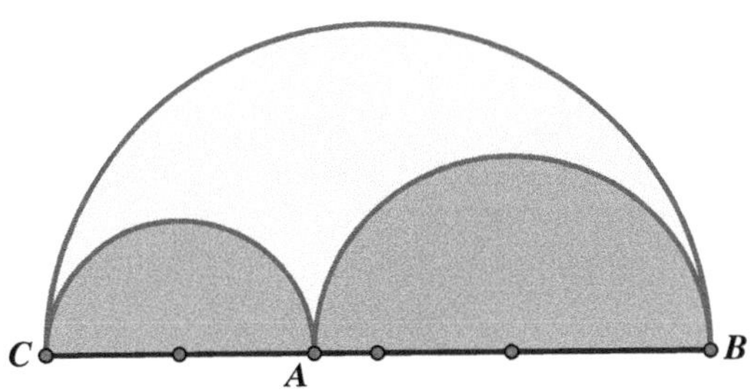

Figure 4-38

areas of the three semicircles is 2:3:5, we seek the ratio of the areas of the lighter shaded region to the largest semicircle.

Since the ratio of the diameters of the three semicircles is 2:3:5, the ratio of the areas is $2^2:3^2:5^2 = 4:9:25$. For convenience, we can assume that the larger semicircle has an area of 25, which implies that the two smaller semicircles have areas of 4 and 9, respectively. This would indicate that the area of the lighter shaded region has an area of $25 - 13 = 12$. Therefore, the ratio of the area of the lighter shaded region to the area of the largest semicircle is 12:25.

Partitioning a Square into Thirds

Given a square piece of paper and a pencil, the task is to partition the square into thirds without any tools beyond the pencil and the paper.

Begin by taking the square piece of paper and folding it in half as shown in Figure 4-39a. Then draw the diagonal of the square and of one of the rectangles, as shown in Figure 4-39b. We then have $\Delta JIP \sim \Delta APD$, since they have equal vertical angles and alternate interior angles of the parallel lines. Since $JI = \frac{1}{2}AD$, we know that $JP = \frac{1}{2}AP$, which makes $JP = \frac{1}{3}AJ$. Repeat the process with the other rectangle as shown in Figure 4-39c. The same argument can be used to show that $KQ = \frac{1}{3}CK$. A similar argument can be made to show that $CL = \frac{1}{3}AC$ and $AM = \frac{1}{3}AC$.

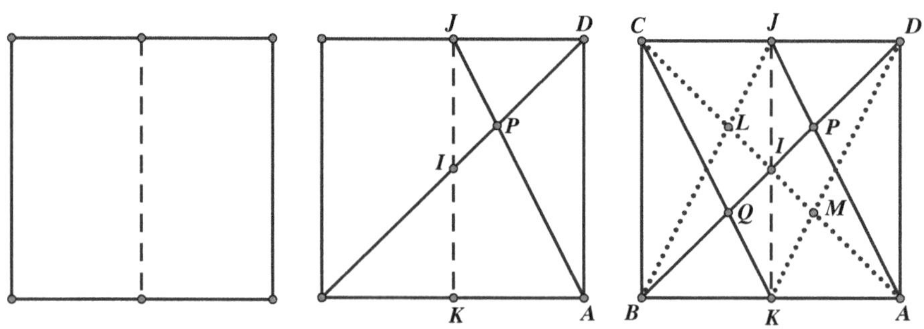

Figure 4-39a Figure 4-39b Figure 4-39c

We then draw lines joining the trisection points *L* and *Q* and the trisection points *P* and *M* to get *ON* and *SR*, as shown in Figure 4-39d, which then partition the square into three equal-area rectangles: *CNOB*, *NRSQ*, and *DRSA*.

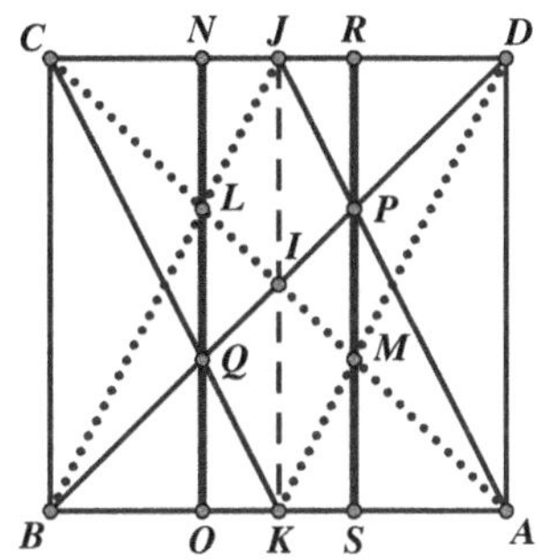

Figure 4-39d

Determining the Size of the Sequential Square

In Figure 4-40, we have three squares lined up so that their vertices at points *D*, *G*, and *H* are collinear. The side length of the smallest square is 7 and the side length of the middle square is 9. We need to find the area of the largest square.

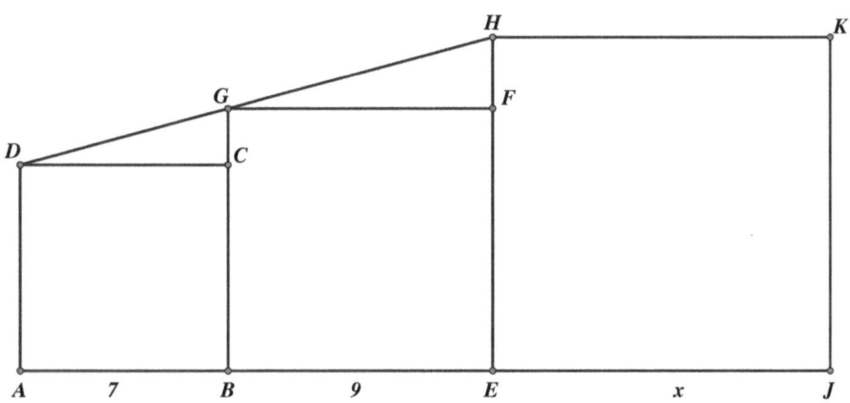

Figure 4-40

As result of the collinearity, we have $\Delta DGC \sim \Delta GHF$. We also know that $GC = GB - AB = 9 - 7 = 2$. We also know from the similarity that $\frac{GC}{DC} = \frac{HF}{GF}$. Therefore, $\frac{2}{7} = \frac{HF}{9}$, and $HF = \frac{18}{7}$, and so, $HE = 9 + \frac{18}{7} = \frac{81}{7}$.

Thus, the area of square

$$HKJE = \left(\frac{81}{7}\right)^2 = \frac{6561}{49} \approx 133.9.$$

A Square Logistics Dilemma

A $6' \times 6'$ square wall needs to be completely covered in paint. Half the wall is painted, and the remaining part is still a square which is 6 feet from top to bottom and 6 feet from side to side. How can this be possible?

This is simply an issue of logistics and explanation. In Figure 4-41, we show a square wall whose sides are 6 feet in length. In the middle, we show a square that is 6 feet from top to bottom and 6 feet from side to side; that is, its vertices are 6 feet apart. The square in the middle has half the area of the large square.

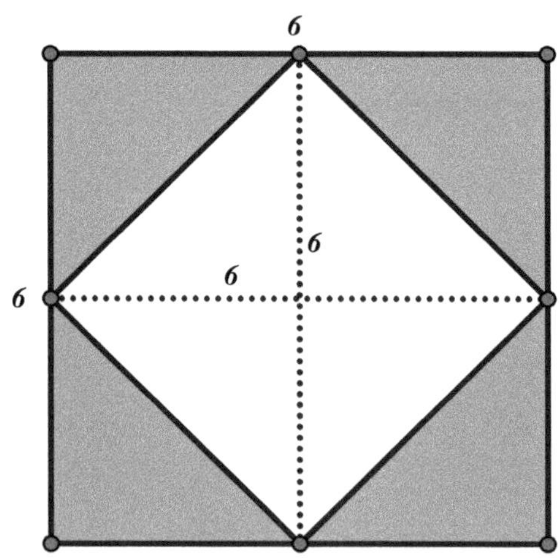

Figure 4-41

The Wonder of Equilateral Triangles

The complete symmetry of an equilateral triangle brings an incredible number of relationships into geometry. We begin with an equilateral triangle constructed on each side of a randomly drawn triangle, such as we see in Figure 4-42, where equilateral triangles *ADC*, *AFB*, and *BEC* are drawn on the sides of triangle *ABC*.

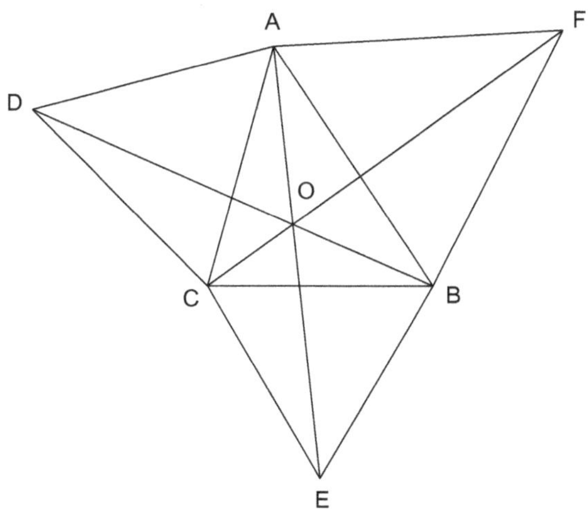

Figure 4-42

Quite surprisingly, the segments joining each vertex of the random triangle to the remote vertex of the equilateral triangle drawn on the opposite side, namely, *AE*, *BD*, and *CF*, are congruent.

What is particularly noteworthy is that this relationship is true for any triangle *ABC*, regardless of its shape. To ensure that the reader is convinced this is true in all cases, we shall prove that it is true. The trick is to identify the proper triangles that need to be proven congruent. They are not easy to identify. One pair of these congruent triangles, $\Delta ACE \cong \Delta BCD$, is shown in Figure 4-43. These two congruent triangles will establish $AE = BD$. The other segments can be proved congruent in a similar way with another pair of embedded congruent triangles. This indicates that all three line segments are equal, namely, $AE = BD = CF$.

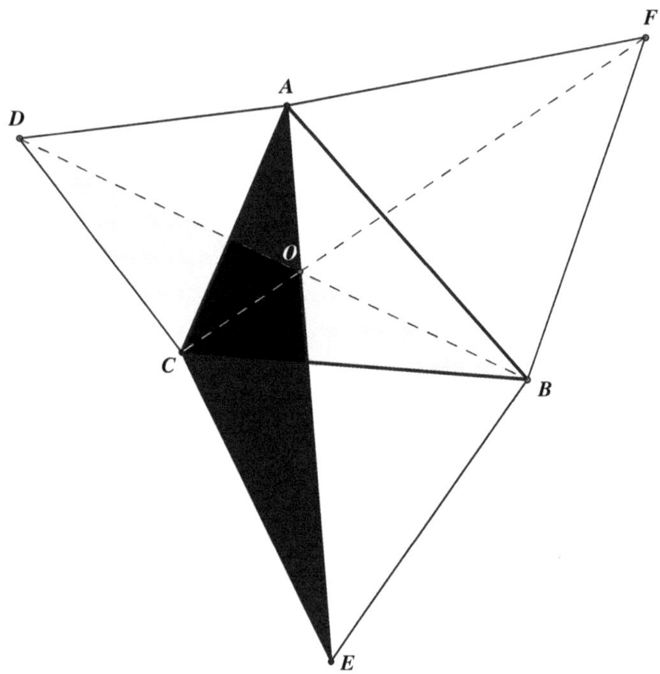

Figure 4-43

Not only is point O a common point for the three segments in Figure 4-42, but it is also the only point in the triangle where the sum of the distances to the vertices of the original triangle is a minimum. This is often called the *minimum-distance point* of the triangle ABC. The sum of the distances from any other point in the triangle to the three vertices would be greater than from this minimum-distance point.

As if that were not enough, point O is the only point in the triangle where the sides subtend equal angles. That is, $\angle AOC = \angle COB = \angle BOA = 120°$, as seen in Figure 4-44.

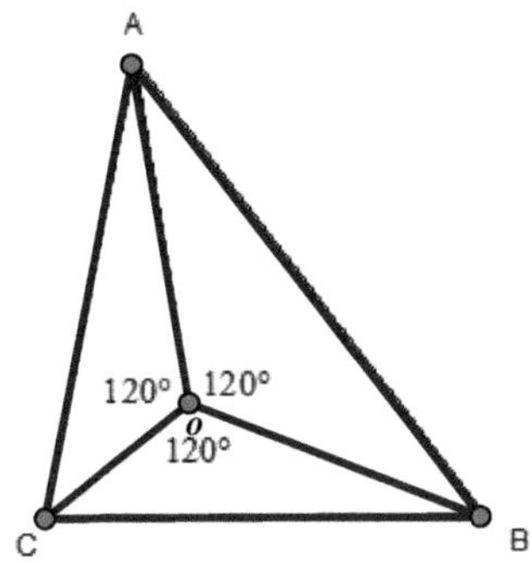

Figure 4-44

When we locate the center of the each of the three equilateral triangles (which can be done in a variety of ways, such as getting the point of intersection of the three altitudes, medians, and angle bisectors) and join these center points, we find that we have created another equilateral triangle, as shown in Figure 4-45.

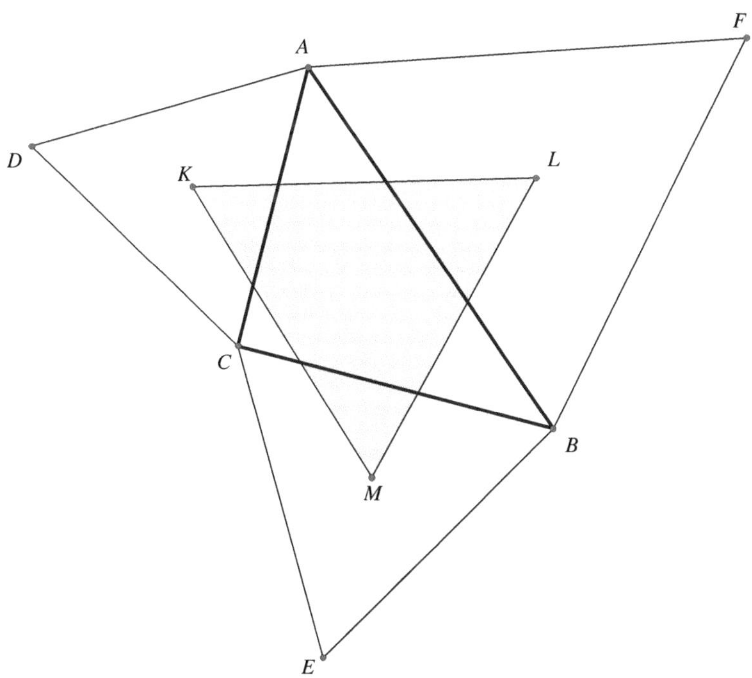

Figure 4-45

Regardless of the shape of the original triangle, the above relationships all hold true. An interesting question you might ask yourself is what would happen if point C was on side AB, thereby collapsing the original triangle, as shown in Figure 4-46.

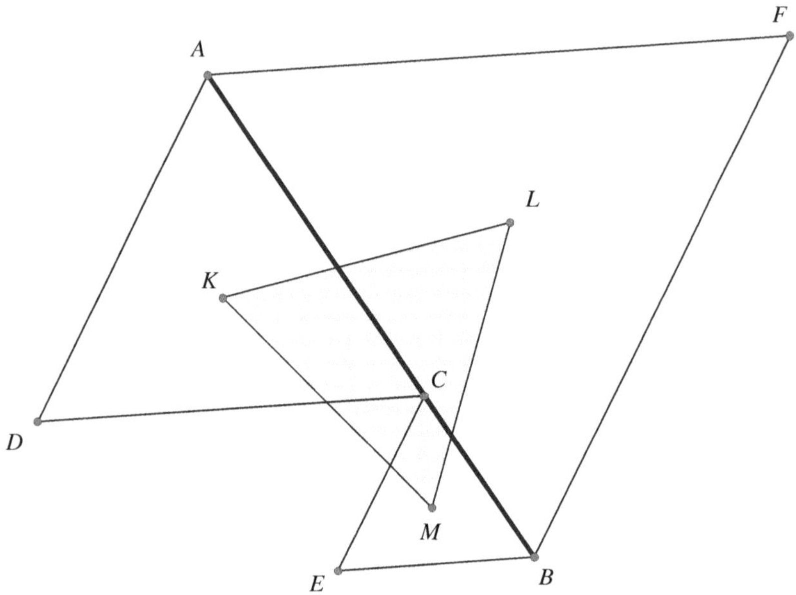

Figure 4-46

Lo and behold, our equilateral triangle is preserved. Perhaps even more astonishing (if anything could be) is the generalization of this theorem. That is, suppose we were to construct similar triangles appropriately placed on the sides of our randomly drawn triangle, and joined their centers (this time we must be consistent as to which "centers" we choose to use—centroid, orthocenter, incenter,[1] etc.).

[1] The *centroid* is the point of intersection of the medians of a triangle. The *orthocenter* is the point of intersection of the altitudes of a triangle. The *incenter* is the point of intersection of the angle bisectors of a triangle. The *circumcenter* is the center of the circumscribed circle.

The resulting figure will be similar to the three similar triangles. Furthermore, we find that the circumcircles of the three equilateral triangles are concurrent at point *O*, as can be seen in Figure 4-47.

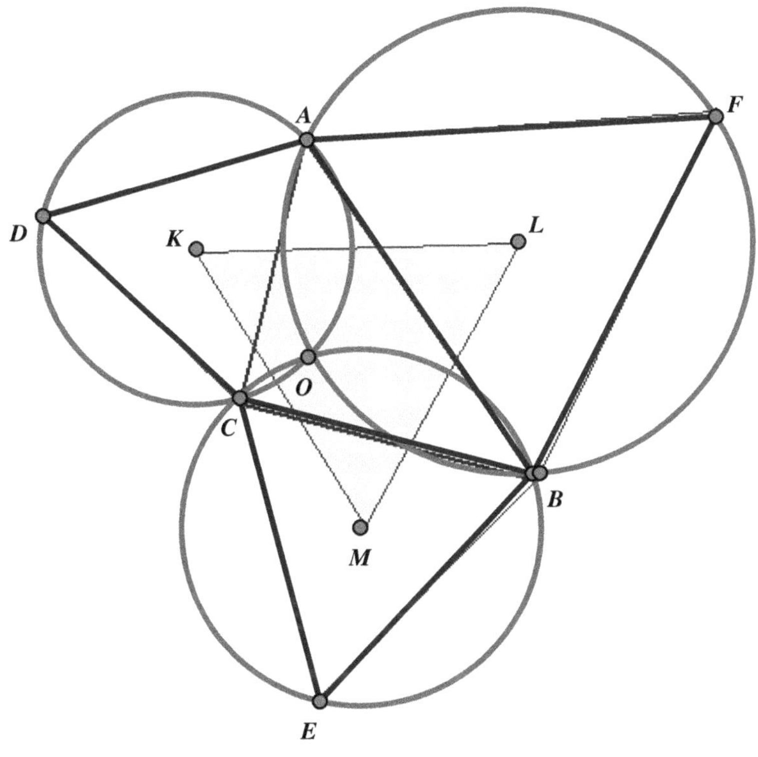

Figure 4-47

The Surprising Pentagram Angles

The Pythagoreans used the regular pentagram as their symbol. The pentagram is also on the American flag 50 times. We know that the sum of the angles of a triangle is 180° and quadrilaterals' angles have a sum of 360°. But what is the sum of the angles of a pentagram? Although easily justifiable, we shall assume that all pentagrams have the same angle sum. Suppose we want to get the angle sum of an arbitrarily-drawn pentagram, such as the one in Figure 4-48.

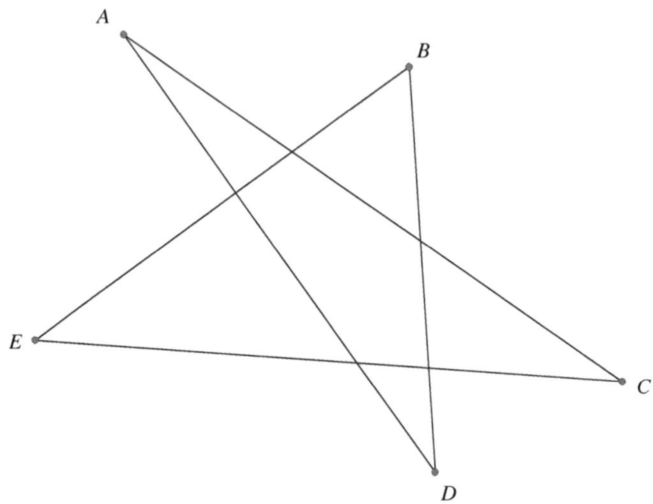

Figure 4-48

Follow the procedure shown in Figure 4-49a–4-49f.

Begin by placing a pencil on *AC* pointing at *A*, and rotating it through $\angle A$ so that it is now on *AD* pointing at *A*.

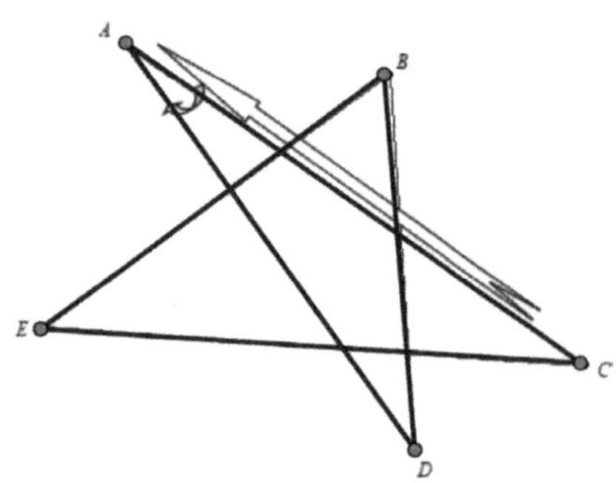

Figure 4-49a

Then rotate the pencil through $\angle D$ so that it is now on BD pointing at B.

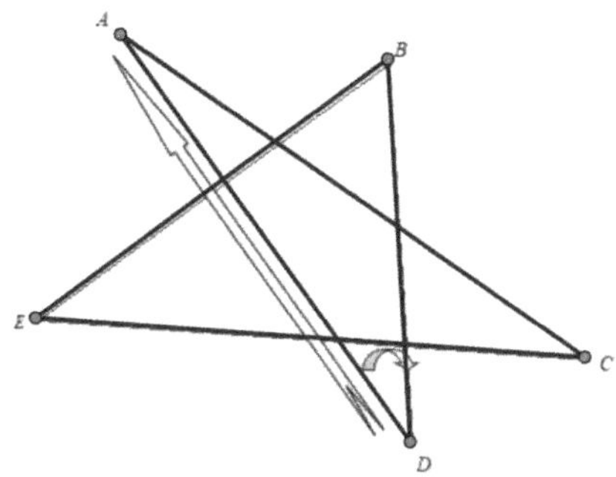

Figure 4-49b

Then rotate it through $\angle B$ so that it is now on BE pointing at B.

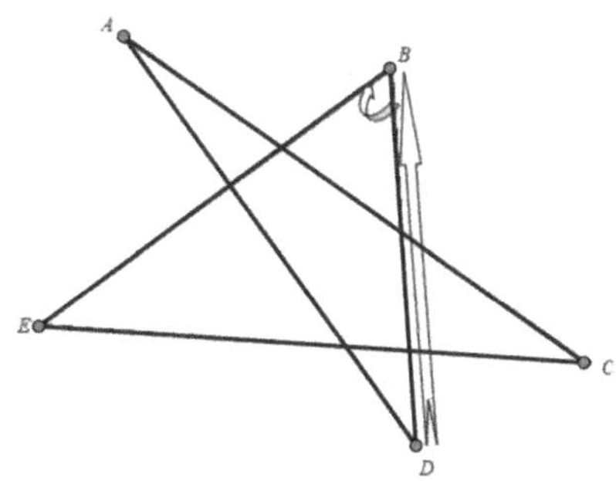

Figure 4-49c

Then rotate it through $\angle E$ so that it is now on EC pointing at C.

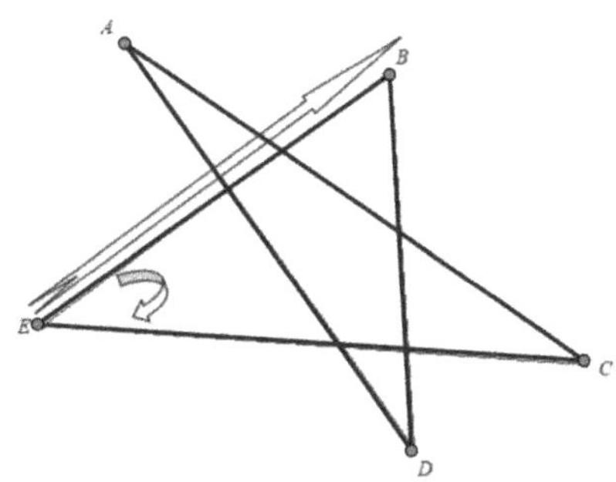

Figure 4-49d

Lastly, rotate it through $\angle C$ so that it is now on AC pointing at C.

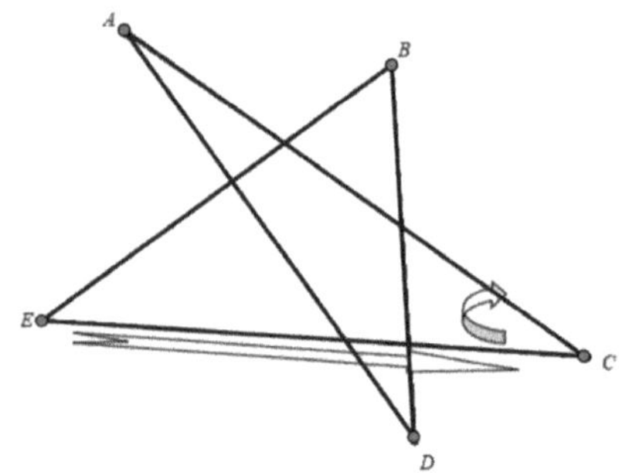

Figure 4-49e

This is the opposite direction of its starting position.

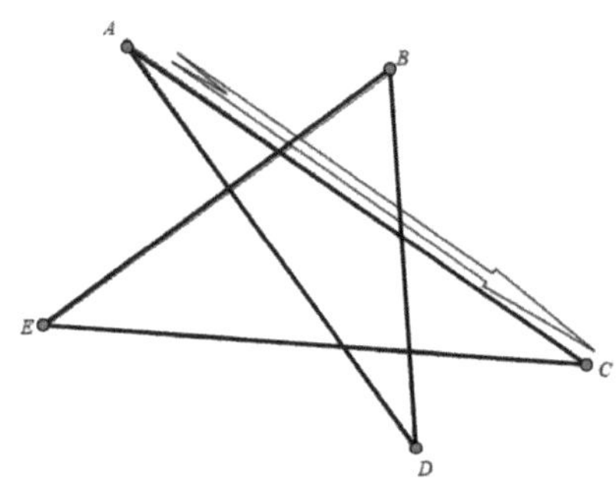

Figure 4-49f

The pencil reversed its direction, which is the same thing as a rotation of 180°, implying that the angle sum of the pentagram (through which the pencil was rotated, angle by angle) is 180°. Again, notice how, through the sequence of angle moves, the pencil's direction changed by 180°.

For those who feel more comfortable with a geometric proof, the following demonstration is provided. Note that we are accepting the notion that the angle sum of the vertices of a pentagram is the same for all pentagrams. Since the type of pentagram was not specified, we can assume the pentagram to be regular, or one that can be inscribed in a circle with all of its vertices lying on the circumference.

In Figure 4-50, we notice that each of the angles is now an inscribed angle of the circle, and so has half the measure of the intercepted arc.

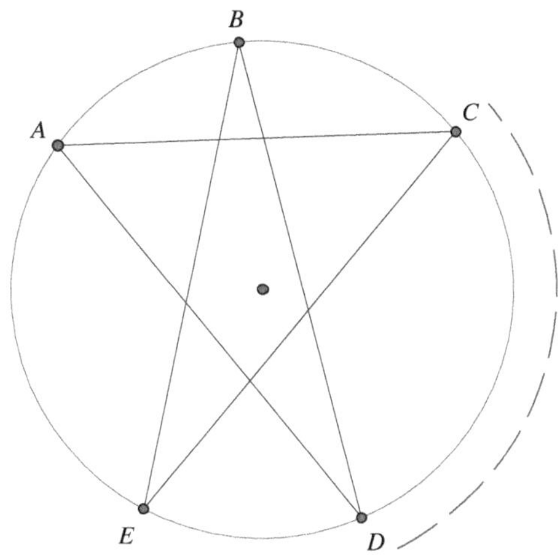

Figure 4-50

Consequently, we obtain the following:

$\angle A = \frac{1}{2}\widehat{CD}$ [this reads "measure of angle A equals one-half the measure of arc CD"]

$$\angle B = \frac{1}{2}\widehat{ED}$$

$$\angle C = \frac{1}{2}\widehat{AE}$$

$$\angle D = \frac{1}{2}\widehat{AB}$$

$$\angle E = \frac{1}{2}\widehat{BC}.$$

If we add these equalities, we obtain

$$\angle A + \angle B + \angle C + \angle D + \angle E = \frac{1}{2}(\widehat{CD} + \widehat{ED} + \widehat{AE} + \widehat{AB} + \widehat{BC}) = \frac{1}{2} \cdot 360° = 180°.$$

That is, the sum of the measures of the angles of the pentagram is one-half the degree measure of the circle, or 180°. Note that there was no loss of generality by allowing the non-specified pentagram to assume a more useful configuration. In other words, we were able to change the figure to one that is more manageable and lose nothing in the process.

A Geometric Conundrum Is Deceptively Difficult or Not

On the surface, the conundrum presented here seems rather simple. There is a clever solution if you can see the trick, but most people trained in high school geometry are apt to pass it by. Yet once the clever solution is shown the conundrum becomes quite simple, disappointing those who missed seeing the solution right from the start. So here it is. Try it without looking at the second figure, as that will give away the clever solution.

In the Figure 4-51, point E lies on AB and point C lies on FG. The area of parallelogram $ABCD = 20$ square units. The challenge is to find the area of parallelogram $EFGD$. Typically, the area of a parallelogram is found by taking the product of its height and base lengths. This is not possible here, so another procedure must be found.

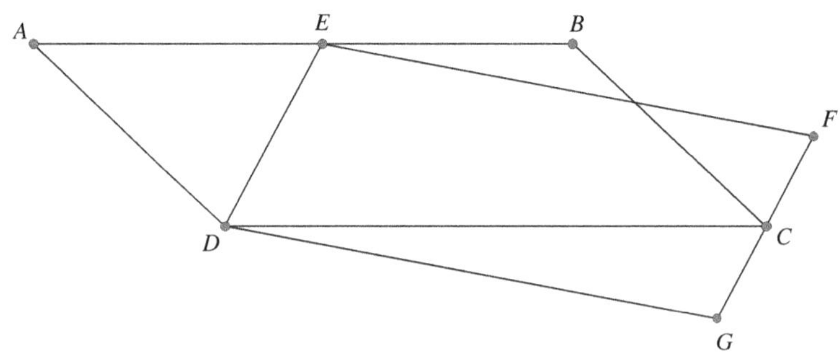

Figure 4-51

Although the solution would not occur to many people at first thought, the problem can be readily solved using the tools found in a high school geometry course. Begin by drawing *EC* as in Figure 4-52.

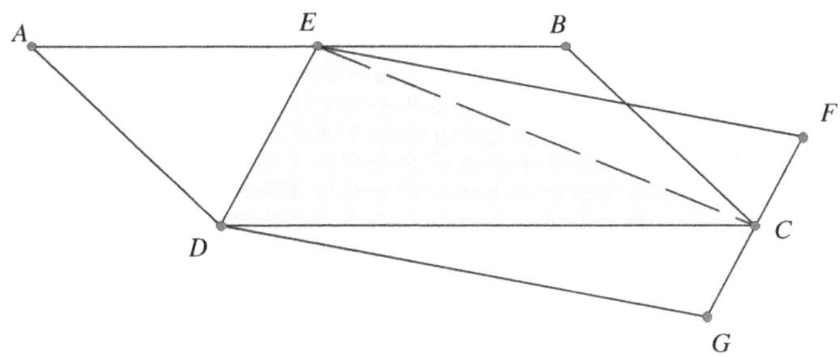

Figure 4-52

Since triangle *EDC* and parallelogram *ABCD* share a common base (*DC*) and a common altitude (a perpendicular from *E* to *DC*), the area of triangle *EDC* is equal to one-half the area of parallelogram *ABCD*.

Similarly, since triangle *EDC* and parallelogram *EFGD* share the same base (*ED*), and the same altitude to that base (a perpendicular from *C* to *ED*), the area of triangle *EDC* equals one-half the area of parallelogram *EFGD*.

Now, since the area of parallelogram *ABCD* and the area of parallelogram *EFGD* are both equal to twice the area of triangle *EDC*, the areas of the two parallelograms must be equal. Thus, the area of parallelogram *EFGD* equals 20 square units.

Although the method that we have just shown is not often used, it is effective and efficient. Nevertheless, this problem can be solved quite elegantly by solving a simpler analogous problem (without loss of generality). The original given conditions were that the two parallelograms had to have a common vertex (*D*), and a vertex of one had to be on the side of the other, as shown with points *E* and *C*. Now, let us suppose that *C* coincided with *G*, and *E* coincided with *A*.

This satisfies the conditions of the original problem and makes the two parallelograms coincide. Thus, the area of parallelogram $EFGD = 20$ square units.

This last solution uses extremes. That is, we might consider point E to be on the extreme end AB, such as on point A. Similarly, we could place C on point G and satisfy all the conditions of the original problem. Thus, the problem is trivial, in that the two parallelograms overlap. Using extremes without loss of generality can be a very powerful solution process.

Recognizing Perfect Geometric Shapes

Triangles are considered "perfect" if their perimeter is numerically equal to the area. There are two right triangles that fit this definition of perfection. The right triangle whose sides are 6, 8, and 10 has an area and perimeter numerically equal to 24. Also, the right triangle who sides are 5, 12, and 13 has an area and perimeter equal to 30.

There also are three perfect obtuse scalene triangles. The obtuse triangle whose sides are 9, 10, and 17 has a numerical area and perimeter equal to 36. To get the perimeter, we just need to add $9+10+17=36$. One method to find the area of this triangle is Heron's formula, which is $\sqrt{s(s-a)(s-b)(s-c)}$ for a triangle whose sides have lengths a, b, and c, and $s=\frac{1}{2}(a+b+c)$. For this perfect triangle, the area is

$$\sqrt{18(18-9)(18-10)(18-17)}=\sqrt{18(9)(8)(1)}=36,$$

which is numerically equal to the perimeter.

The triangle whose sides are 6, 25, and 29 has an area and perimeter numerically equal to 60. An obtuse triangle whose sides are lengths 7, 15, and 20 has an area and perimeter numerically equal to 42. Thus, we have identified what are believed to be the only perfect triangles.

We have identified only two perfect rectangles, those whose areas are numerically equal to their perimeters. One of the rectangles has a length of 6 and width of 3. Therefore, its area is $3 \times 6 = 18$, and its perimeter is $3 + 3 + 6 + 6 = 18$. The other perfect rectangle is actually a square of side length 4. Its area is $4 \times 4 = 16$, and its perimeter is $4 + 4 + 4 + 4 = 16$.

Only Lines Solve the Problem

Consider the situation where we need to draw a perpendicular line from a point outside a circle to the diameter of the circle, as shown in Figure 4-53. However, we are only able to use straight lines and no other tools such as a pair of compasses. The problem just requires a knowledge of high school geometry.

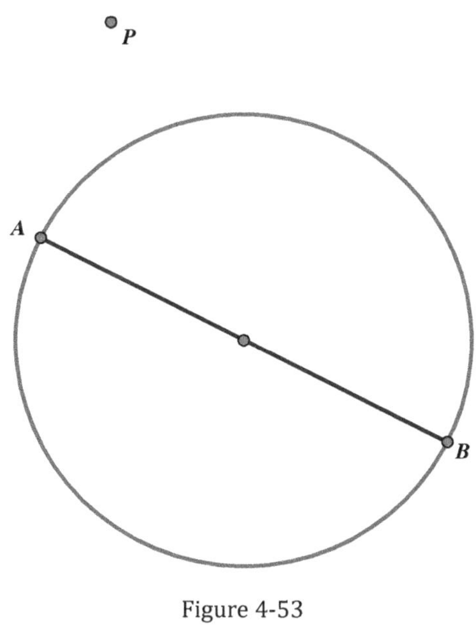

Figure 4-53

The solution requires us to use only straight lines that can be connected between two given points. We begin by connecting point *P* to the endpoints A and B of diameter AB, intersecting the circle at points C and D, respectively, as can be seen in Figure 4-54. We then draw lines DA and BC intersecting at point E. When we draw the line PE, it turns out to be the required perpendicular we sought from the start, as it is perpendicular to BA extended at point G. The justification for this perpendicularity is that $\angle ADB = 90° = \angle ACB$ since they are both inscribed in a semicircle defined by the diameter AB. Since the altitudes of a triangle are concurrent, we know that BAG must be an altitude of triangle PEB, and therefore, $BG \perp PE$. The noteworthy thing is that we did the construction with only an unmarked straightedge.

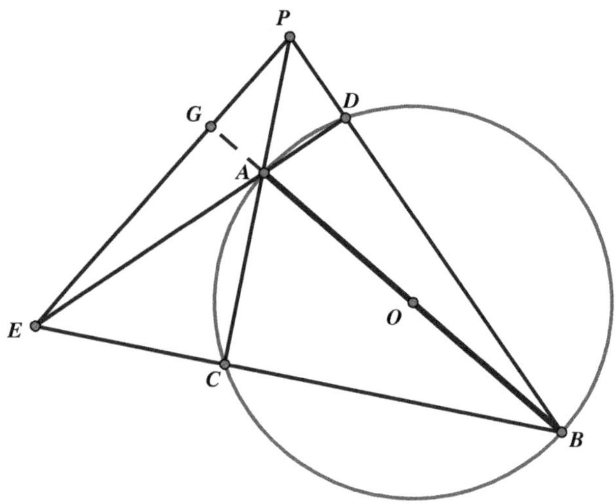

Figure 4-54

Ceva's Very Helpful Relationship

Here, we address a geometry issue that is too frequently neglected. In many cases, the concurrency of lines is taken for granted. We just "know" that the altitudes of a triangle are concurrent, meaning they contain a common point of intersection. Similarly, we often take for granted that a triangle's medians or angle bisectors are concurrent. The topic of concurrency of lines in a triangle deserves more attention than it usually gets in an elementary geometry course. In order to put these assumptions to rest, we can establish an extremely useful relationship. This will be done with the help of the famous theorem first published in *De Lineis Se Invicem Secantibus Statica Construction* (Milan, 1678) by the Italian mathematician Giovanni Ceva (1647–1734), and which now bears his name. This relationship is really quite simple and makes a usually difficult topic—concurrency—almost trivial! It is included here because it's such a valuable tool in geometry.

In simple terms, Ceva's theorem says that if you have three concurrent line segments (*AL*, *BM*, and *CN)* joining a vertex of a triangle with a point on the opposite side, then the product of the alternate segments along the sides (determined by the concurrent segments) are equal. The converse of this theorem tells us that if the products of the alternating segments are equal, then the lines determining these segments are concurrent. This may appear at first to be a bit complicated, but note in Figure 4-55 how the products of the alternate segments along the sides of the triangle are equal: $ANBLCM = NBLCMA$.

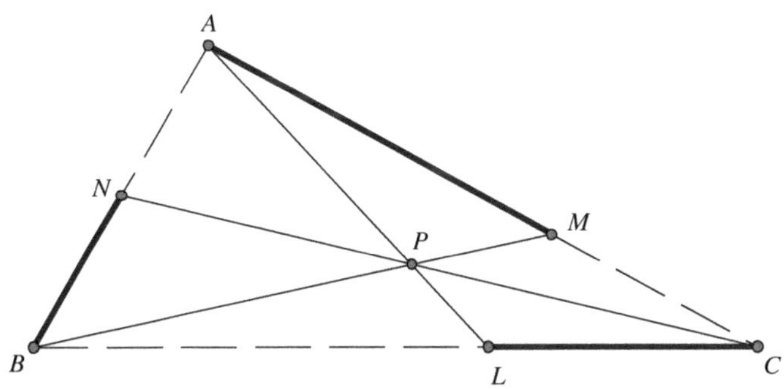

Figure 4-55

The theorem can be more formally stated as follows:

Ceva's Theorem: *The three lines containing the vertices A, B, and C of* ΔABC *and intersecting the opposite sides at points L, M, and N, respectively, are concurrent if, and only if,* $\frac{AN}{NB} \cdot \frac{BL}{LC} \cdot \frac{CM}{MA} = 1$, or $ANBLCM = NBLCMA$.

Figure 4-56 shows how the point of intersection could be inside or outside the triangle.

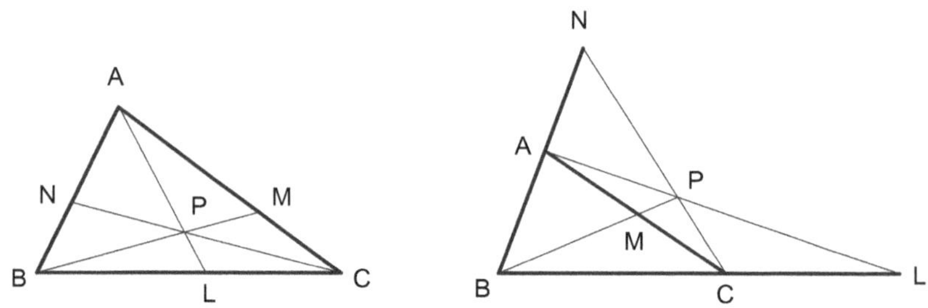

Figure 4-56

It is easier to follow the proof with the left side of the diagram, and to verify the validity of the theorem with the right side of the diagram. In any case, the statements made in the proof hold true for *both* diagrams.

Ceva's theorem is an equivalence (or biconditional) and therefore requires two proofs, one the converse of the other. We shall first prove that *if the three lines containing the vertices of ΔABC and intersecting the opposite sides in points L, M, and N, respectively, are concurrent, then* $\frac{AN}{NB}\cdot\frac{BL}{LC}\cdot\frac{CM}{MA}=1$.

The proof is offered here for the curious and ambitious.

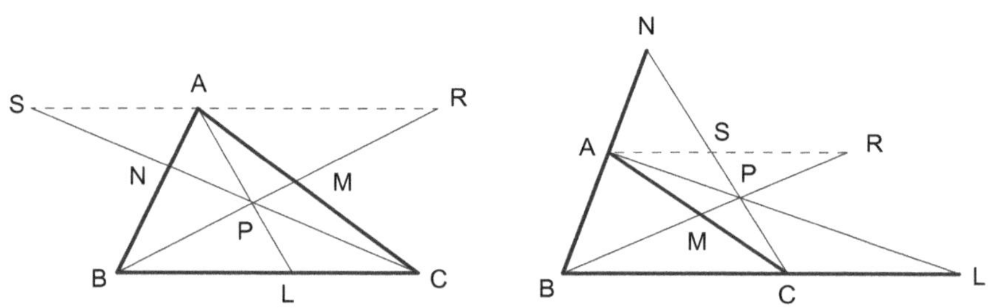

Figure 4-57

Proof: In Figure 4-57, we begin by drawing a line containing *A* and parallel to *BC* that intersects *CP* (extended) at *S* and *BP* (extended) at R. The parallel lines enable us to establish the following pairs of similar triangles:

$$\Delta AMR \sim \Delta CMB\text{, therefore } \frac{AM}{MC} = \frac{AR}{CB}. \tag{I}$$

$$\Delta BNC \sim \Delta ANS\text{, therefore } \frac{BN}{NA} = \frac{CB}{SA}. \tag{II}$$

$$\Delta CLP \sim \Delta SAP\text{, therefore } \frac{CL}{SA} = \frac{LP}{AP}. \tag{III}$$

$$\Delta BLP \sim \Delta RAP\text{, therefore } \frac{BL}{RA} = \frac{LP}{AP}. \tag{IV}$$

From (III) and (IV) we get:

$$\frac{CL}{SA} = \frac{BL}{RA}.$$

This can be rewritten as:

$$\frac{CL}{BL} = \frac{SA}{RA}. \tag{V}$$

Now by multiplying (I), (II), and (V), we obtain the desired result:

$$\frac{AM}{MC} \cdot \frac{BN}{NA} \cdot \frac{CL}{BL} = \frac{AR}{CB} \cdot \frac{CB}{SA} \cdot \frac{SA}{RA} = 1.$$

We rearrange the terms and invert the fractions to get:

$$\frac{NA}{BN} \cdot \frac{BL}{CL} \cdot \frac{MC}{AM} = 1.$$

An Unexpected Concurrency

A fascinating and unexpected point of concurrency in a triangle was first established by the French mathematician Joseph-Diaz Gergonne (1771–1859). Gergonne reserved a distinct place in the history of mathematics as the founder of the first purely mathematical journal, *Annales des mathématiques pures et appliqués*, in 1810. However, his fame in mathematics is largely due to the discovery of the relationship between a triangle and its inscribed circle: The line segments joining the vertices of the triangle with the three points of tangency with the circle are concurrent.

To prove this relationship involving the concurrency of lines in a triangle, we can use Ceva's theorem, which we have just considered earlier.

Gergonne's Theorem: *The lines connecting a vertex of a triangle to the opposite side's point of tangency with the inscribed circle of the opposite side are concurrent.* This point of concurrency is known as the *Gergonne point* of the triangle.

We see this relationship in Figure 4-58, with the vertices of triangle ABC connected to the three points of tangency L, M, and N of the inscribed circle.

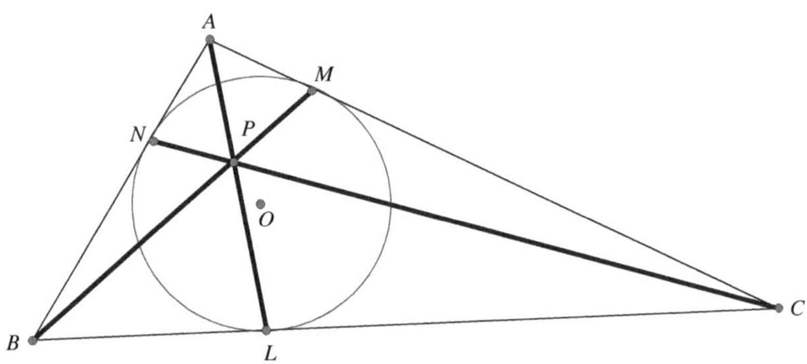

Figure 4-58

Proof: Circle O is tangent to sides AB, AC, and BC at points N, M and L, respectively. It follows that $AN = AM$, $BL = BN$, and $CM = CL$. Each of these equalities may be written as $\frac{AN}{AM} = 1$, $\frac{BL}{BN} = 1$, and $\frac{CM}{CL} = 1$.

By multiplying these three fractions we get $\frac{AN}{AM} \cdot \frac{BL}{BN} \cdot \frac{CM}{CL} = 1$. Therefore, $\frac{AN}{BN} \cdot \frac{BL}{CL} \cdot \frac{CM}{AM} = 1$, which, as a result of Ceva's theorem, implies that AL, BM, and CN are concurrent. This point is the Gergonne point of ΔABC.

Neat and (relatively) simple! Yet this fact is actually not well known. These easy-to-understand relationships make geometry fun.

Determining the Area Between Two Concentric Circles

In Figure 4-59, we have two concentric circles with point O as the center. The distance from the point of tangency of the inner circle P to the circumference of the outer circle R is 1 unit long. The challenge is to find the area between the two circles.

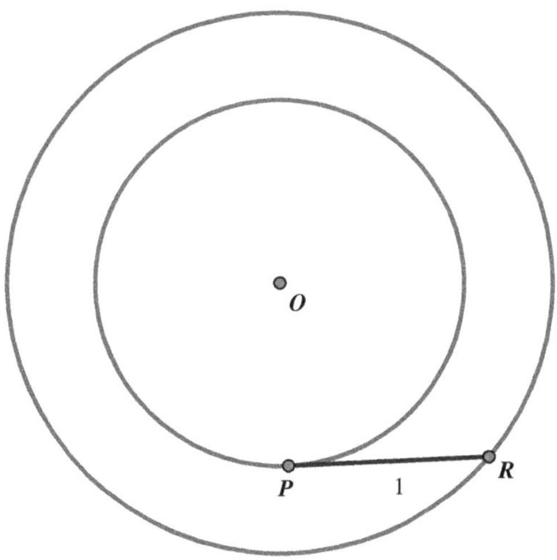

Figure 4-59

To find the area between the two circles, we need to establish the radii of each of the two circles, which are $OP = a$ and $OR = b$, as shown in Figure 4-60. Since the radius drawn to the point of tangency is perpendicular to the tangent, we have right triangle *POR*. We apply the Pythagorean theorem to get $a^2 + 1^2 = b^2$, or $b^2 - a^2 = 1$. To get the area between the circles, we subtract the areas of the two circles: $\pi b^2 - \pi a^2 = \pi(b^2 - a^2)$. However, we just established that $b^2 - a^2 = 1$. Therefore, the area between the circles is simply π.

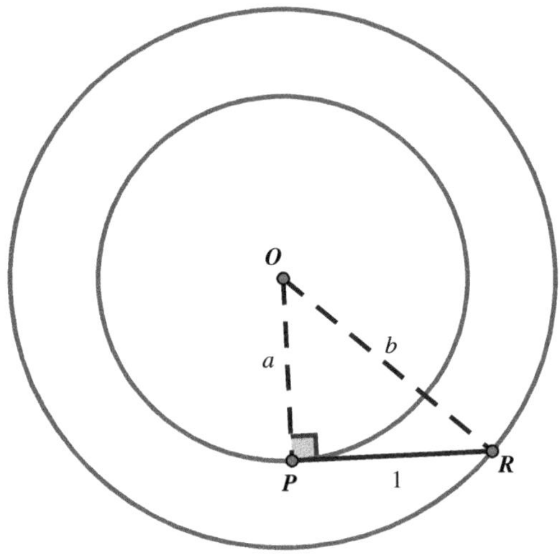

Figure 4-60

Fascinating Equilateral Triangle Relations

In Figure 4-61, we find one equilateral triangle inscribed in another equilateral triangle in such a way that the vertices of the inscribed equilateral triangle are all at trisection points of the sides of the larger equilateral triangle. What percent of the area of the larger equilateral triangle is occupied by the smaller equilateral triangle?

A traditional solution applies to each of the two triangles the formula for the area of an equilateral triangle: $\frac{s^2\sqrt{3}}{4}$, where s is the length of the side. However, a more elegant solution begins by realizing that *BD* partitions the original triangle into thirds. That is, in Figure 4-62, we have $\Delta ABD = \frac{1}{3}\Delta ABC$. Similarly, in triangle *ABD*, *DE* is a

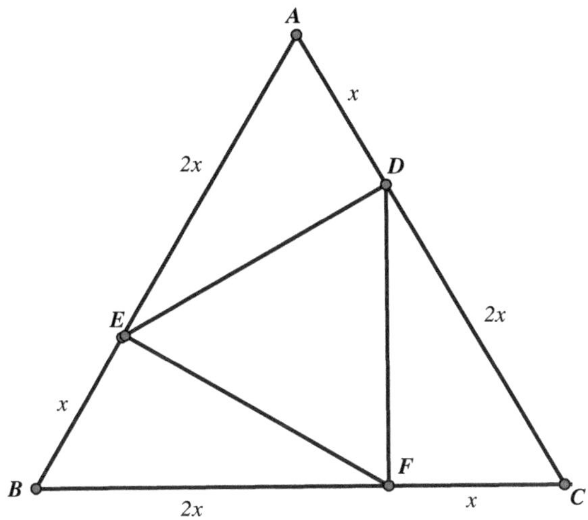

Figure 4-61

trisection line, so that $\Delta ADE = \frac{2}{3}\Delta ABD$. Therefore, $\Delta ADE = \frac{2}{9}\Delta ABC$. We can easily prove that $\Delta ADE \cong \Delta DCF \cong \Delta BEF$, so that the areas $\Delta ADE + \Delta DCF + \Delta BEF = 3\left(\frac{2}{9}\right)\Delta ABC = \frac{2}{3}\Delta ABC$. That leaves $\Delta DEF = \frac{1}{3}\Delta ABC$, which answers our challenge.

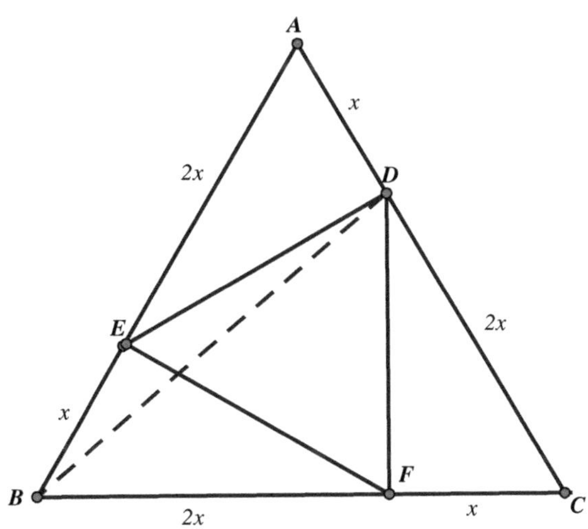

Figure 4-62

The Unexpected Isosceles Triangle

When a perpendicular is drawn to the base of an isosceles triangle to intersect one side and an extension of the other side, a second isosceles triangle is formed. In Figure 4-63, line *EFD* is perpendicular to base *BC* of triangle isosceles *ABC* and intersects sides *AC* and *AB* at points *F* and *E*, respectively. Surprisingly, triangle *EAF* is also an isosceles triangle.

This can be justified by noting that in right triangle *BDE* we have $\angle E + \angle B = 90°$, and in right triangle *CDF* we have $\angle CFD + \angle C = 90°$. Since the base angles of isosceles triangle *ABC* are equal, that is, $\angle B = \angle C$, and vertical angles $\angle CFD = \angle E$, we thus we have $\angle AFE = \angle E$, which then determines that triangle *AEF* is isosceles.

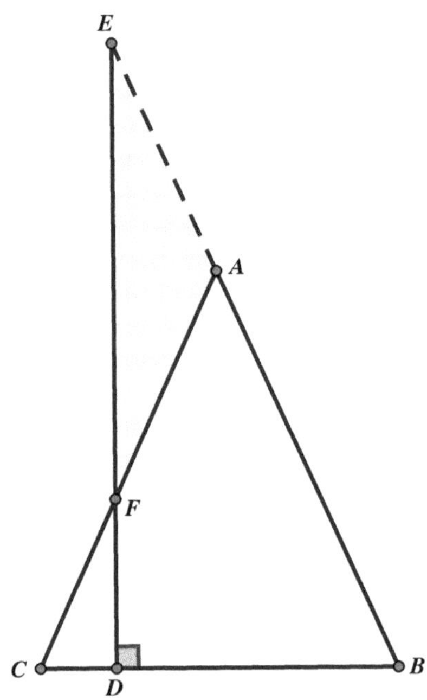

Figure 4-63

The Trick of the Unexpected Concurrency

Draw a circle and then inscribe any random quadrilateral in that circle. We come up with four lines that will all go through one point—a quite unexpected concurrency. First, locate the midpoints of each of the four sides of the quadrilateral. Then construct a line from the midpoint of each of the sides of the quadrilateral that is perpendicular to the opposite side. In Figure 4-64, we have drawn perpendiculars from the midpoints *K, L, M,* and *N* of the sides *AB, BC, CD,* and *AD,* respectively, to the opposite sides of the quadrilateral. Quite unexpectedly, it turns out that these four lines are all concurrent at point *E.*

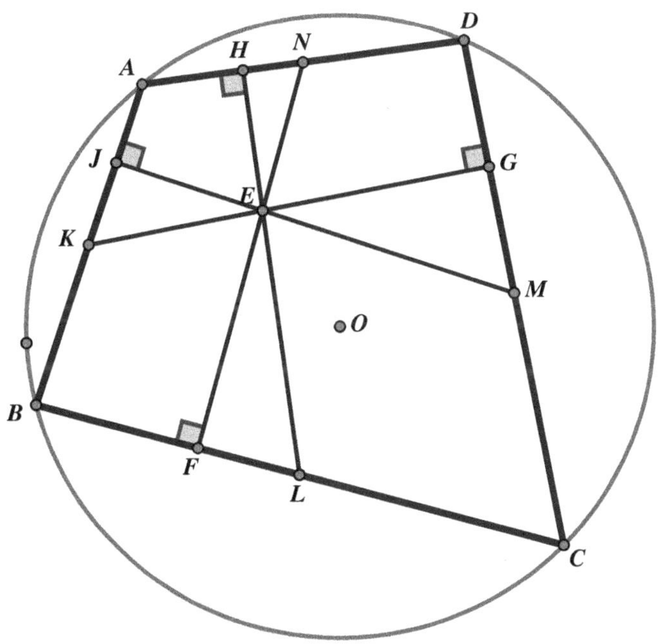

Figure 4-64

In Figure 4-65, we have added line OL, which is perpendicular to the line BC. We will now consider triangles POL and PEN. The lines joining midpoints of the opposite sides of the cyclic quadrilateral bisect each other at point P, and therefore, $PL = PN$. Also, vertical angles $\angle OPL = \angle EPN$. We then extend OP its own length to point E. Since $OL \parallel NF$, we have $\angle PLO = \angle PNE$ and $\Delta POL \cong \Delta PEN$. This establishes that point E is on NF. This can then be repeated for the other perpendiculars, which all contain point E.

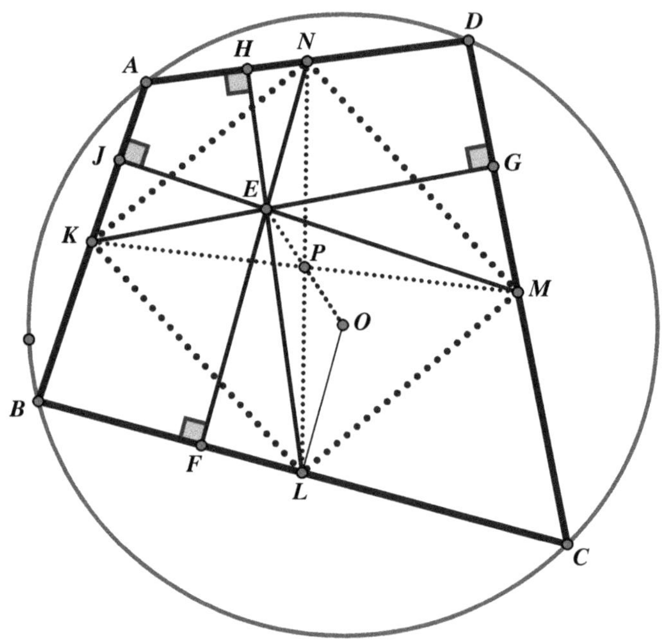

Figure 4-65

The Unexpected Perpendicular Line

Begin by drawing a circle with two perpendicular chords as we show in Figure 4-66, where AC is perpendicular to BD. Then draw the quadrilateral $ABCD$. When a perpendicular line is drawn from point P, the intersection point of the perpendicular diagonals of

ABCD, to *AB*, its extension intersects the midpoint, *F*, of *DC* quite unexpectedly.

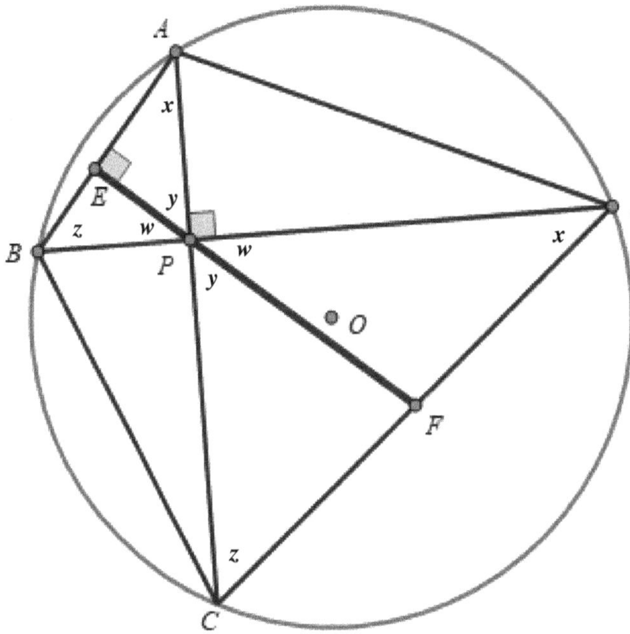

Figure 4-66

This can be easily justified by noting in triangle *APE* we have $x + y = 90°$, and in triangle *BEP* we have $w + z = 90°$ and with vertical angles $\angle APE = \angle FPC = y$. Therefore, since $z = y$, triangle *PFC* is isosceles. Analogously, triangle *PFD* is also isosceles, and therefore, since $CF = PF = DF$, we have established that *F* is the midpoint of *DC*.

The Tricky Line That Divides a Triangle in Half

Draw any triangle and select a point on one side. The trick here is to draw a line from that point to another side so that the triangle is divided into two equal-area regions.

The positioning of this line is somewhat unusual. Follow along as we construct the line and then justify how it divides the triangle into two equal areas. We will call our triangle *ABC*, as shown in Figure 4-67. We will then select a random point *P* on side *BC* and then draw the line *AP*. The next step is to locate the midpoint of *BC*, which we will call point *M*. Through point *M* we will draw a line parallel to *AP* intersecting side *AC* at point *R*. This will now determine the sought-after line *PR*, which we will show divides the original triangle *ABC* into two equal-area regions.

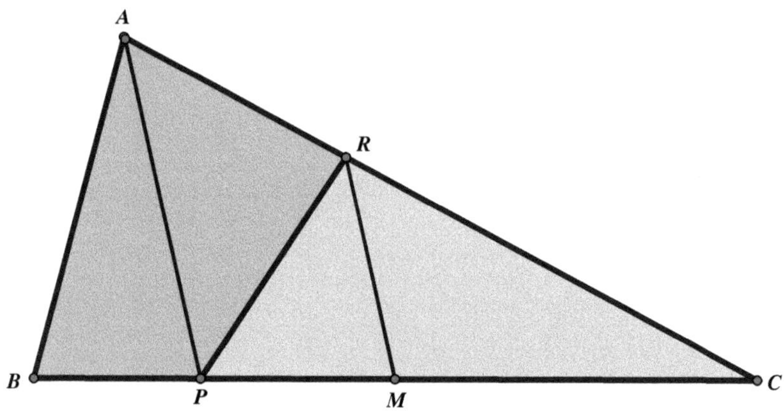

Figure 4-67

As we see in Figure 4-68, draw line *AM*, which divides the triangle *ABC* into two equal-area triangles, namely, triangle *ABM* and triangle *ACM*. The reason that they are equal in area is that they have a common altitude from *A* to *BC*, and they have equal bases since $BM = MC$. Now here comes the tricky part.

Consider the trapezoid *ARMP*, where we can show that triangle *ARP* is equal in area to triangle *AMP*, since they share a common base *AP* and their altitudes lie between parallel lines. Refer now to Figure 4-69, where point *Q* is the intersection of *AM* and *PR*. If we remove triangle *AQP* from the configuration, we are left with two new equal-area triangles, namely, triangle *ARQ* and triangle *PMQ*. When we remove triangle *PQM* from triangle *ABM*, and at the same time add its

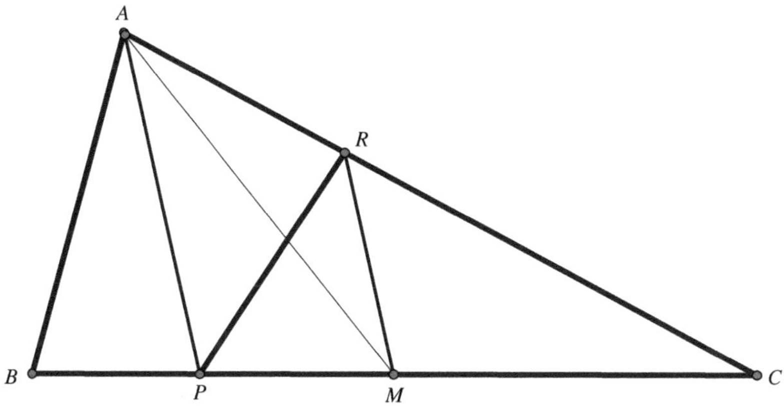

Figure 4-68

equal, namely, triangle *ARQ*, alongside the remaining region of triangle *ABM*, namely, the quadrilateral *AQPB*, we then have established that the area of quadrilateral *ARPB* is equal to the area of triangle *PRC*. This then has line *PR* partition triangle *ABC* into two equal areas, as was our original goal.

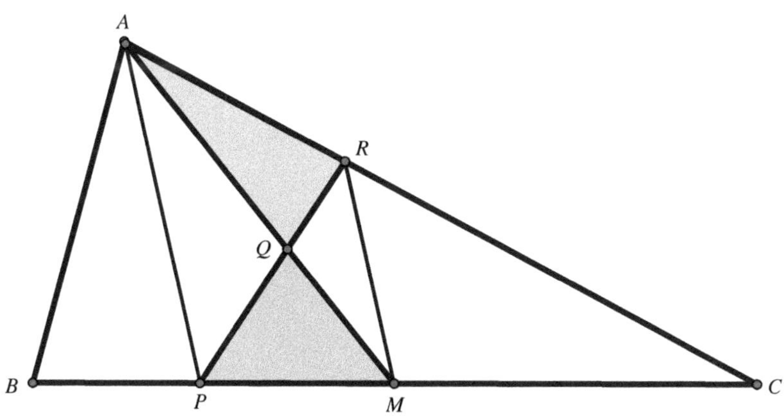

Figure 4-69

The Surprising Trick of the Angle Bisectors of a Quadrilateral

Begin by drawing a quadrilateral of any shape (which is technically called a *general quadrilateral*). Then construct the bisectors of each of the four angles and extend them till each bisector intersects the bisectors of the quadrilateral's adjacent angles. In Figure 4-70, we show the general quadrilateral *ABCD* with each of the four angle bisectors drawn so that they intersect adjacent angle bisectors at the points *E, F, G,* and *J*, which, amazingly, all lie the same circle. Any three noncollinear points will always lie on a unique circle; however, four noncollinear points will only in special cases lie on the same circle. They are known as *concyclic points* – a rather special arrangement.

To justify this phenomenon, we must show that the opposite angles of quadrilateral *EFGJ* are supplementary, that is, have a sum of 180°. To do that, we need to focus on ΔBCG and ΔADE. Both triangles

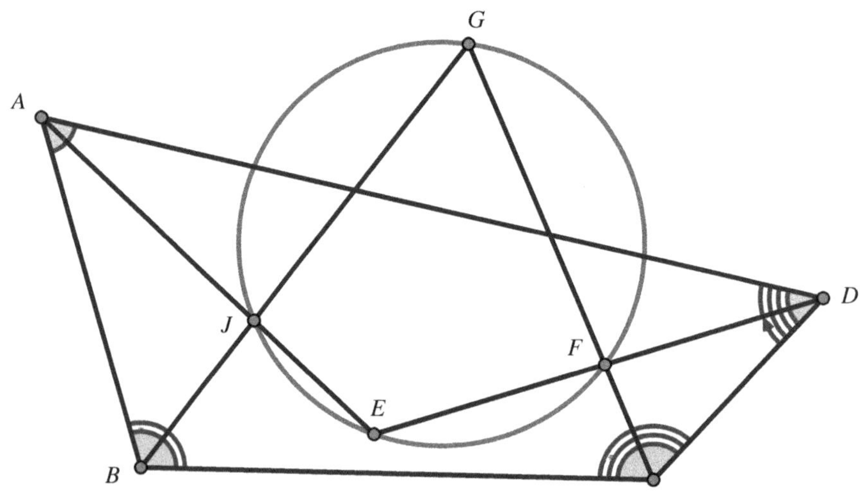

Figure 4-70

have two of the angles whose total combined sum equals 180°, that is, $\angle GBC + \angle GCB + \angle EAD + \angle EDA = 180°$, since the sum of the angles of quadrilateral *ABCD* is 360°. Since the two triangles ΔBCG and ΔADE have an angle sum of 360°, that leaves $\angle G + \angle E = 180°$, which then establishes that quadrilateral *EFGJ* is a cyclic quadrilateral, since the opposite angles are supplementary.

The Angle Bisectors of a Parallelogram Trick

We can take the previous trick one step further by beginning with a parallelogram, rather than with a general quadrilateral, and then drawing the four angle bisectors as before. Here, the bisectors' intersections create a rectangle formed by the four concyclic points *E*, *F*, *G*, and *J*, as shown in Figure 4-71. We can use the argument from the previous example by noting that consecutive angles of a parallelogram are supplementary. That indicates that in ΔBCG we have $\angle GBC + \angle GCB = \frac{1}{2} \cdot 180° = 90°$. The same argument can be used with the other angles of the rectangle *EFGJ*.

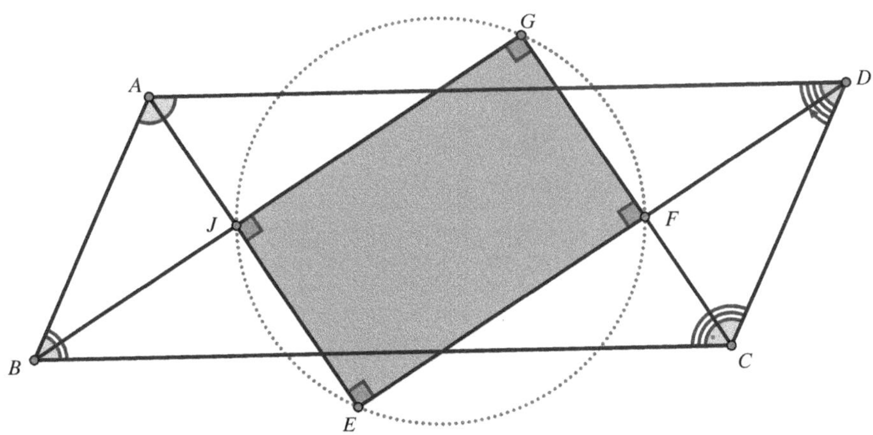

Figure 4-71

The Surprising Appearance of a Rectangle

In geometry, it is often necessary to use some construction tools, or at least to have a good eye for drawing geometric shapes and bisecting lines and angles. Begin with a circle and then choose any four points on the circle that will give some form of random quadrilateral—that is, a quadrilateral that has no equal sides, equal angles, or parallel sides. Consider the quadrilateral *ABCD* inscribed in circle *O*, shown in Figure 4-72.

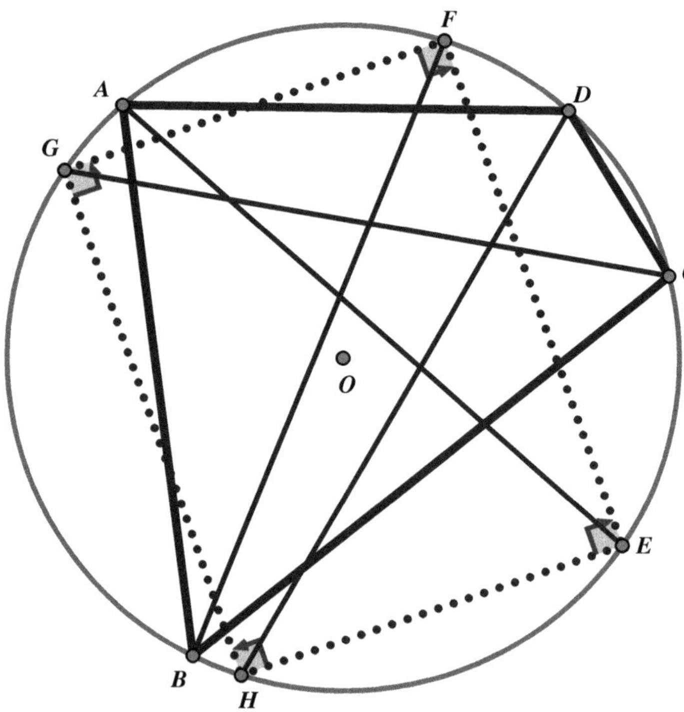

Figure 4-72

Begin by bisecting each of the angles of the quadrilateral to intersect the opposite side of the circle. Regardless of the shape of the original quadrilateral *ABCD*, the four points of contact on the circle of the angle bisectors, when joined consecutively, will yield a rectangle, which is *EFGH*. Quite surprising!

This can be justified by noting that the angle bisector of a triangle intersects the perpendicular bisector of the opposite side of the triangle at the circumcircle of the triangle. In Figure 4-73, focus on ΔABD, where *KE*, the perpendicular bisector of side *BD* is also a diameter of the circle *O*. This determines the right angle *EFG*. This can be repeated at points *E*, *H*, and *G*. Thus, the result is rectangle *EFGH*.

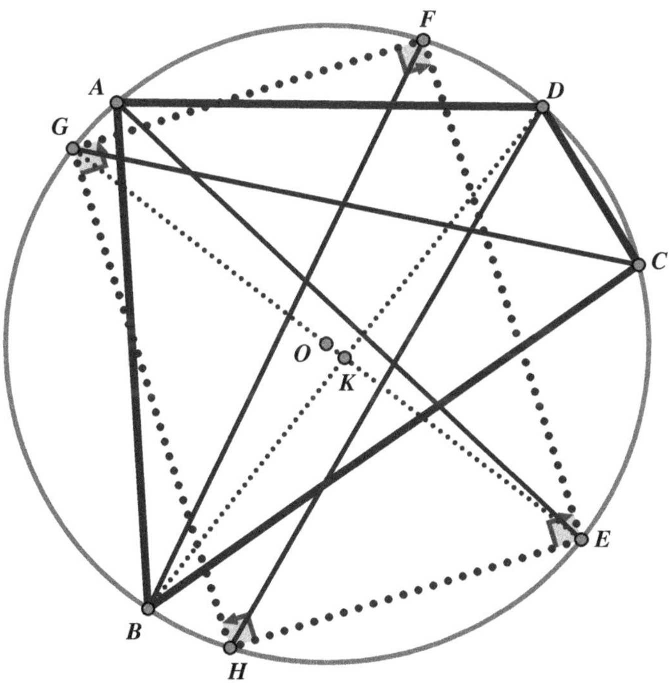

Figure 4-73

You may want to take this one step further and focus on points marked *W*, *X*, *Y*, and *Z*, shown in Figure 4-74, and be amazed to see that they will also lie on a unique circle.

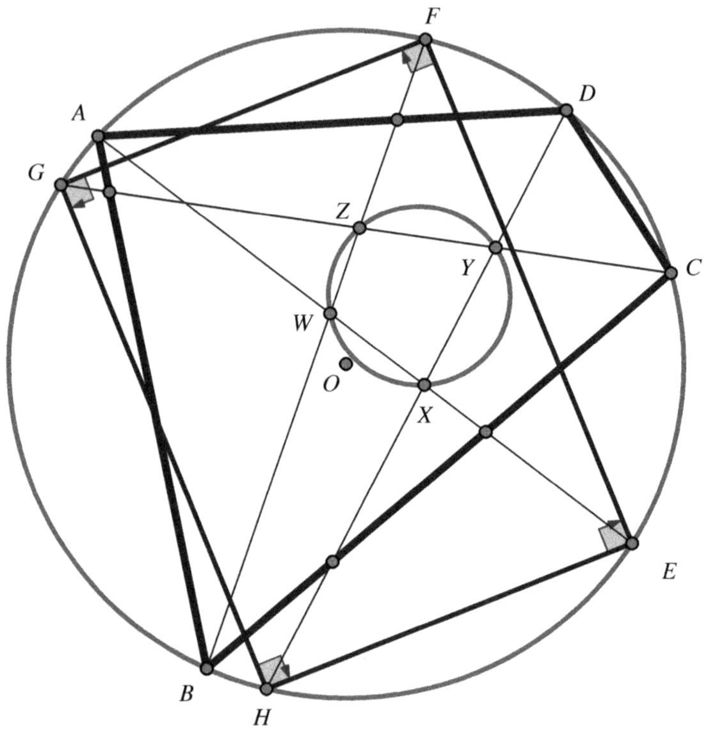

Figure 4-74

As the sum of the angles of a quadrilateral is 360°, half the angle sum is 180°. This angle sum can be displayed as $\angle WAB + \angle ABW + \angle YDC + \angle DCY = 180°$. Since the angle sum of triangles *AWB* and *DYC* is 360°, when we subtract the angle sum $\angle WAB + \angle ABW + \angle YDC + \angle DCY = 180°$ from this we find that $\angle AWB + \angle DYC = 180°$. Therefore, the vertical angles $\angle ZWX + \angle ZYX = 180°$, which makes quadrilateral *WXYZ* cyclic. Put another way, the points *W*, *X*, *Y*, and *Z* lie on the same circle.

Chapter 5

Mathematics Tricks

Creating tricks in mathematics is entertaining, enriching, and educational. It can be quite impressive to discover and exhibit such tricks. Determining why the tricks work can be quite a revealing experience. This chapter is devoted to tricks with numbers. They have been selected because they provide the reader with some unusual aspects of mathematics that can be easily demonstrated. Enjoy!

Simply, a Cute Trick

Here is a cute little trick that one can play at the dinner table, since it can easily be done mentally. Have your audience select a number from 1 to 10 and keep it secret. Ask them to multiply their secret number by 9. Then ask them to add the digits of the product and subtract 5 from the sum. Still without divulging any information, have them identify the letter of the alphabet that corresponds with that number (A = 1, B = 2, C = 3, etc.) Have them select a country in Europe whose name begins with that letter. Then ask them to select an animal whose name begins with the last letter of the European country. You will be able to tell them that the animal that they have selected is a kangaroo. They will surely be amazed at your talent as a trickster.

Here is the reason for your hidden "talent." When they multiply their original number by 9 and add the digits of the product, the result will be 9. Subtracting 5 from 9 yields 4. The fourth letter of the alphabet is a D. The only country in Europe whose name begins with a D is Denmark. The only *popular* animal whose name begins with the last letter of the word Denmark, a K, is the kangaroo. And so, you have used a few tricks to impress your audience.

The 777 Number Trick

Once again, the nature of numbers enables us to create an impressive trick. Have your friend select any number between 500 and 1000; then have him or her add 777 to this number. If the sum exceeds 999, have him or her remove the thousands digit and add it to the units digit. Then ask your friend to subtract that sum from the number originally selected. You can astonish your friend by predicting that the result is 222.

To secure an understanding of the procedure, let's try one example now. We select the number 600. We add to it 777: $600 + 777 = 1377$. The thousands digit, 1, is then removed and added to the units digit to get $377 + 1 = 378$. Subtracting these two numbers (the original number and the one just obtained) $600 - 378 =$ **222.**

You may wonder why this trick works as it does. The sum of 777 and any number between 500 and 1000 will always exceed 1000. Dropping the 1 from the thousands place and adding it to the units digit is tantamount to subtracting 999 from the number. That is, $-999 = -1000 + 1$. If we represent the selected number as n, then what is being done is: $n - (n + 777 - 999) = n - n - 777 + 999 = 222$.

Suppose we used a number other than 777 as our "magic" number, perhaps the number 591. Then our friend would have ended up with 408 every time, regardless of which number he chose between 500 and 1000. For a "magic" number of 733, the end result will always be 266. Remember, the selected number cannot be less than 500 or else you will not get a sum in the thousands. At the same time, the selected number should not be greater than 999, or you might get a 2 in the thousands place, which would also ruin this scheme.

When a Quotient Equals a Difference

We now offer the trickster the ability to challenge the audience to find two numbers whose quotient and difference are both equal to 5. Typically, the audience will immediately start speculating with various integers to see which ones suit the challenge. Unfortunately, that would be the wrong approach. When the quotient of two numbers is 5, then their difference will be four times the smaller of the two numbers. For example, the quotient of $30 \div 6 = 5$, and the difference between $30 - 6 = 24$, which is 4×6. One possible pair of numbers that satisfies this relationship would be $\frac{25}{4}$ and $\frac{5}{4}$. The difference of these two numbers is 5, as is their quotient. In general, when $x - y = \frac{x}{y} = a$, we have $x = \frac{a^2}{a-1}$ and $y = \frac{a}{a-1}$. Another pair of numbers that satisfies this relationship is $\frac{144}{11}$ and $\frac{12}{11}$. Their difference is $\frac{132}{11} = 12$, and their quotient is $\frac{144}{11} \div \frac{12}{11} = \frac{144}{11} \cdot \frac{11}{12} = 12$. Can you find other such numbers?

The Trick of Showing That 5 = 4

This trick will surely have your audience guessing. It might also make them a bit uncomfortable until you expose the secret behind the trick. To do the trick, we are given that $a + b = c$. Certainly, it is clearly true that $4a + 4b - 5c = 4a + 4b - 5c$.

When we add these two equations, we get the following:

$$5a + 5b - 5c = 4a + 4b - 4c.$$

By factoring each side of the equation, we get:

$$5(a + b - c) = 4(a + b - c).$$

When we divide both sides by the common factor $(a + b - c)$, we get $5 = 4$.

Pause to allow your audience to marvel over this ridiculous result. You can then expose the error, namely, that division by 0 is forbidden in mathematics. Since we began with the understanding that $a + b = c$, it follows that $(a + b - c) = 0$. Therefore, we divided both sides by 0, which leaves us with a ridiculous result. This further justifies why division by 0 is forbidden in mathematics.

Deciphering a Strange Sequence of Numbers

The challenge here is to determine the next number in the sequence:

181, 920, 212, 223, 242, 526, 272, 829...

This is actually a very simple challenge once you notice that the numbers presented are separated incorrectly. Notice that if we consider these as two-digit numbers starting with 18, we get the following sequence:

18, 19, 20, 21, 22, 23, 24, 25, 26, 27, 28, 29.

The next numbers would be 30, 31, 32, which then creates the following sequence:

18, 19, 20, 21, 22, 23, 24, 25, 26, 27, 28, 29, 30, 31, 32.

And therefore, the sought-after next terms of the given sequence of numbers would be the following sequence:

181, 920, 212, 223, 242, 526, 272, 829, 303, 132.

Impress with a Surprise

When asked to divide 59 by 10, we get 5 with a remainder of 9. Ask your audience to find a number which when divided by 10 leaves a remainder of 9, and when divided by 9 leaves a remainder of 8. This number will also leave a remainder of 7 when divided by 8. This pattern will continue such that when this number is divided by 3 it will leave a remainder of 2, and finally, when divided by 2 it will leave a remainder of 1. The challenge for your audience is to find a number that has these characteristics.

One such number is 14,622,042,959. However, it is unrealistic to expect the audience to come up with this number. Therefore, there must be a small number you might want to have your audience search for, such as 3,628,799, which also satisfies the original challenge. Yet, the smallest solution can be obtained by looking for the least common multiple of the numbers 1, 2, 3, 4, ..., 8, 9, 10, which is $2^3 \times 3^2 \times 5 \times 7 = 2520$, and then subtract 1 to get 2519, which is the smallest number that satisfies the criteria requested.

Maximizing a Product

Here is a simple problem to see if your audience can think logically. They are challenged to find the largest product formed by two five-digit numbers, where no digit is used more than once across the two numbers. Logical thinking will prevail if they realize that the largest numerals must be at the left end of each number. The desired product will be: $86{,}420 \times 97{,}531 = 8{,}428{,}629{,}020$. Notice that the two factors use all 10 digits exactly once. You should also recognize the pattern of how these two numbers were developed to create the largest product. Hint: Each of the two numbers have alternating digits in descending order.

The Determine Your Deleted Number Trick

The interaction between a trickster and a player might follow the dialogue shown in Figure 5-1.

The Trickster Says:	**The Player:**
Select any five-digit number (not one with five of the same digits) and keep it secret.	35,630
Rearrange the digits of your selected number in any way you wish.	53,306
Subtract these two numbers.	53,306 – 35,630 = 17,767
Delete any one digit (1 – 9) of this difference, *but not 0*, and keep it secret.	17,6~~7~~6
Rearrange the digits of this number and tell me the resulting number.	7, 6, 1, 6
I can tell you which digit you removed. It was a 7.	Mystified!

Figure 5-1

The player tends to try this a few times only to find that the trickster can always somehow determine which number has been removed. How is this possible? Yes, some knowledge of number properties will help you figure out this secret.

The important fact to know here is that the difference of two numbers with the same digits will always have a digit sum that is a multiple of 9, such as 9, 18, 27, 36, 45, etc. So, when the trickster is told which digits remain after the player deleted one digit, the trickster simply takes the sum of the digits and selects the digit that would bring that digit sum to the next multiple of 9. That must be the digit which was removed. So, you can see that this "trick" is completely dependent on the notion that the difference of two numbers with the same digit sum must have a digit sum which is a multiple of 9. Such arithmetic trivia gives one a better grasp of number properties.

Guessing-Your-Number Trick

This is a simple and quick number trick to impress your friends at dinner while waiting for the food to be served. Have your friend select any three-digit number. Then have your friend reverse the digits and subtract the two numbers. Now ask your friend to tell you the units digit of this final result, and you will then be able to determine the entire resulting number.

For example, suppose your friend selects the number 673. Reversing the digits and subtracting gets $673 - 376 = 297$. When your friend tells you the units digit is 7, you should be able to reply that the hundreds digit is 2 and the tens digit is 9. The way you do this is that you know that the units digit and the hundreds digit must add up to 9 and the tens digit will always be a 9. Therefore, since our exposed units digit was 7, the hundreds digit had to be 2 and the tens digit will always be a 9. Hence, we ended up with 297. Naturally, you could have asked your friend to expose the hundreds digit and that would allow you to get the number as well, since similarly the sum of the units and the hundreds digit must always be 9.

A Simple Guess-the-Number Trick

Tell your friend to think of a three-digit number where all the digits are different. You and she will do a calculation which will end up with the number she originally selected and did not divulge to you.

Have her do the following calculations with the secretly selected number:

- Double the number and add 4.
- Multiply the result of the previous step by 5 and add 12.
- Multiply the previous result by 10 and subtract 320.
- Delete the last two digits of the previous calculation, and that will be her secret number.

To clarify, we provide a specific example with the secret number 418:

- Double the number and add 4.

$$2 \times 418 + 4 = 840.$$

- Multiply the previous step by 5 and add 12.

$$5 \times 840 + 12 = 4212.$$

- Multiply the previous result by 10 and subtract 320.

$$4212 \times 10 - 320 = 41{,}800.$$

- Delete the last two digits of the previous calculation and that will be her secret number: 418. Surprise!

Another Guessing-Your-Number Trick

Your friend may not need a piece of paper for this trick that uses small numbers. Ask your friend to select a number, double the number, then add 5, and then multiply this result by 5. When your friend tells you the results of his calculation, all you need to do is delete the units digit and then subtract 2, and you will have the number that your friend originally selected.

When we try this trick with the number 36, we double the number to get 72, add 5 to get 77, and multiply this by 5 to get 385. When we get this result, we delete the units digit to get 38, and when we subtract 2, we get the secret number, 36.

Yet Another Guessing-Your-Number Trick

Begin this trick by having your friend select any number and keep it secret. Ask your friend to multiply the number by 4, then take half of the result and multiply that by 7. Your friend then divulges his results, which you divide by 14 to get your friend's secret number.

To make this a bit clearer, we can try this trick with the secret number 28, which we multiply by 4 to get 112, which we divide in half to get 56, which is then multiplied by 7 to get 392. With this number given to us, we divide by 14, which allows us to get the initial number, 28.

A New Guess-Your-Number Trick

Once again, we ask our friend to select a secret number and double it. The friend is then asked to select any *even* number and add it to this result. Taking half of this newly gotten number, you instruct your friend to multiply that by 4. The friend is then asked to subtract from this result double the even number chosen earlier and divulge this new result to you. In order to guess the original secret number, all you need to do is divide this newly exposed number by 4.

When we practice this trick with the number 23, we double the number to get 46, and we select a random even number, say 18, which we add to the 46 to get 64. We then multiply number $64 \times 4 = 256$, take one-half of 256 to get 128, and subtract from that twice our earlier selected even number 18, so that we have $128 - 36 = 92$. This number is then exposed to the trickster, who simply divides it by 4 to get the original secret number, 23.

An Additional Guess-Your-Number Trick

This trick will require somewhat more extensive calculation, since you will be asking your friend to multiply his or her secret number by 50 and then adding 72 to that number. From this result your friend is to subtract 111 and then add 39. This result is then to be divided by 5. Once this last number is exposed to you, you will be ready to do a simple calculation, namely, to take one-tenth of that number, and you will have exposed your friend's secret number.

We can try this trick by beginning with the number 7, which we then multiply by 50, which is 350, and then add 72 to get 422. From this result, we subtract 111 to get 311 and then add 39 to get 350, which divide by 5 to get 70. As the trickster, all that is needed is to take one-tenth of that number, which allows us to expose the secret number of 7.

Further Guessing-Your-Number Trick

This trick is quite unusual, since it requires your friend to perform various operations with her secret number. She will tell you what calculation she does each step without divulging the secret number or the result of each of these calculations.

Perhaps the best way to explain this trick is to use a prototype. Suppose your friend selects the number 8 as her secret number. She is told to multiply the secret number by any number she chooses and to tell you only what that multiplier is, but not the result. Once again, she can multiply that result by any other randomly selected number, which she divulges without revealing the result of the calculation. She then can divide the result by any number she chooses, telling you the divisor but not the resulting quotient. Now she can either choose to multiply or divide the result by another number, which she chooses and again tells you what that number is and which arithmetic process she used, but again not the result of that calculation. This can be done any number of times using multiplication and division in any combination, yet each time telling you just the multiplier or divisor, but not the result of the calculation. When she stops, she will divulge the result of these calculations to you, the trickster.

All you have to do to get the original secret number is to do each of her calculations (with her divisors and multipliers), which you should have listed in order, but begin your calculation with the number 1. When you divide your friend's result by yours, you will have the secret number.

A simulation of this would probably be helpful to get a good grasp of the trick. Remember, we can use any combination of multiplications and divisions and only divulge the number we select as divisor or multiplier. We can begin with the secret number 3 and then *multiply* it by 7 to get 21. We tell the trickster that our first calculation was multiplying by 7. Next, we will *multiply* our result, 21, by 5 to get 105, and again, tell the trickster we just did a multiplication by 5. Now we will choose to do a division, and we will *divide* our result by 2 to get 52.5. We then choose to do another *multiplication*, this time by 5, to get the number 262.5. We can continue this combination of multiplications and division as long as we wish, but each time divulging the operator and the operation. At the end, we need reveal the end result, in this case, 262.5.

In order for us to get the original number, we need to do all the same calculations but beginning with the number 1. Our first calculation was to *multiply* the secret number by 7, which in our case would yield the number 7. The next calculation was to *multiply* by 5, which results in the number 35. We then *divided* by 2, and when we do it, we get 17.5. The next calculation was to *multiply* by 5 again, so we get 87.5. We chose to stop at this point and our result was 262.5. Now all we need to do to get the secret number is divide 262.5 by 87.5 to get 3. How's that for a surprising trick?

A Curious Guessing-Your-Number Trick

Here is a simple number trick that requires work by the trickster and by the friend on whom the trick is being played. As trickster, have your friend select in secret any two-digit number. Have him add 3 to that number, and also subtract 3 from that original number, writing both results on a piece of paper so that you cannot see either. The next step is to square both numbers and then subtract the two squares. Your friend then tells you the result. Now you need to do some calculation to tell him what the original number was.

As trickster, you now need to do is divide the number your friend has given you by 12 to get the secret number.

Let's play the trick out in order to fully appreciate its effect. To do that, we will ask the friend to select a secret number. Suppose she selects the number 52, and keeps it a secret from you, the trickster. You now tell your friend to subtract 3 from her number to get 49, and to add 3 to her number to get 55. By squaring these two numbers she will get 2401 and 3025, respectively. Taking the difference of these two numbers: $3025 - 2401 = 624$. This is the number your friend now divulges to you. As the trickster, you need to do some calculation in order to discover your friend's secret number. Divide the number which your friend gave you, 624, by 12 to get 52, which is the secret sought-after number.

A Guess-the-Number Trick

Here is a number trick that you can play with a friend. Select any number less than 99, write it down on a slip of paper without showing him, and placing it in his pocket. You selected the number 53, but kept it a secret. Tell him to write down any number between 50 and 100, without letting you know what it is. Assume he selected 67. Now subtract the number you have written down on a piece of paper (53) from 99, which is 46, and ask your friend to add this number to his number, that is, $46 + 67 = 113$. Then tell your friend to remove the left-most digit, which in this case is 1, so that he is left with 13. He should then add the deleted digit 1 to the remaining number 13 so that he has the number 14. Now tell him to subtract this number from his original number, that is, $67 - 14 = 53$. He now looks in his pocket, pulls out slip of paper on which you wrote the number 53 and is in total amazement.

This trick can also be done with large numbers. To better follow the procedure, use the table in Figure 5-2.

You write down the number between 100 and 200, say, 167, on a piece of paper in secret.	**167**
Ask your friend to write down a number between 200 and 1000, say, 354.	354
Subtract the number you wrote on the piece of paper from 999 to get 832, and tell your friend to add this number to his previously selected number.	354 + 832 = 1186
Tell your friend to delete the left-most digit, which is 1.	186
Then he is to add that digit to the remaining number.	186 + 1 = 187
Then ask your friend to subtract this result from his original number, which was 354. And he discovers the secret number that you originally wrote on the slip of paper.	354 – 187 = **167**

Figure 5-2

The same technique can be used to determine someone's age over 10 years old. Let's say you want to discover a 36-year-old's age. Have her add 90 to her age, which in this case adds to 126. Then delete the left-most digit, 1, to get 26, and then add this to 26 to 27. Then simply add 9 to this number to get 36, which is the sought-after age.

Another Guess-the-Number Trick

We will present this consider the number trick in steps, as shown in Figure 5-3.

Select any number.	436
Add the next consecutive number.	436 + 437 = 873
Add 9 to this number.	873 + 9 = 882
Divide this by 2.	441
Subtract the original number (436).	441 – 436 = **5**
The result will always be 5.	

Figure 5-3

The trick depends on the number added in the third step, which in this case was 9. If you add 15, then the resulting number will always be 8. This is determined by adding any odd number, adding 1, and dividing it by 2 to get the trick's resulting number. For example, if you chose the number 19 as the adding number, then $19 + 1 = 20$, so the resulting number in that trick will be $20 \div 2 = 10$.

A Final Guess-Your-Number Trick

The trickster asks the player to select any four-digit number. The player has to add the digits and write down that sum. The trickster then asks the player to delete one digit from the original four-digit number. The trickster will at the end determine which number was deleted. From this newly-formed three-digit number, the player should subtract the digit sum from the original four-digit number. This number is then revealed to the trickster, who quickly takes the sum of these digits and subtracts it from the next higher multiple of nine. That result is the digit that was deleted.

Here is an example of how that would work. The player selects the number 1738 and then writes down the number 19, which is the sum of the digits of his original number. The player chooses to eliminate the number 7 from the original number, thus leaving the three-digit number 138, from which the earlier digit sum (19) is to be subtracted to get 119. The trickster then takes the sum of these digits, namely 11, and subtracts it from the next higher multiple of 9, which is 18, to get the number 7, which was the digit originally removed by the player. This should surely amaze the player!

A Trick to Guess Your Friend's Age

For this trick we have a small stipulation, namely, that we need to assume that your friend's age is more than 10. Have your friend add 90 to her age and then remove the hundreds digit and then add it to the remaining number. When your friend tells you this last number, you merely add 9 to get her age.

Let's try this with your friend who is 48 years old, but does not divulge it to you. First, she adds 90 to get 138. She removes the hundreds digit (1) and adds it to the remaining number (38) to get 39. When you are told that the remaining number is 39, you merely add 9 to get 48, which is the correct age!

Guessing a Child's Number

This little trick might be easier to do with a calculator. Have a child choose a two-digit number and which he will keep it secret from you. First, ask the child to square this number. Then the child should take that the original two-digit number, add 8, and square that number as well. Then the child is to subtract the two squares. Now, you have to do some work. Take the number 8, which you used earlier, and double it to get 16. Ask the child for the number he or she obtained at the end and divide it by 16. Then by subtracting 4, which is half of the fixed number 8, from this result, you will arrive at the original number the child selected.

Perhaps an example might clarify this procedure. Suppose the child secretly selects the number 48 and $48^2 = 2304$. By adding 8 to the original number, he will arrive at 56. Squaring each of this result yields $56^2 = 3136$. Subtracting these two squares yields $3136 - 2304 = 832$. Divide this number by 16 (which is twice the 8 used earlier) to get 52. To get the original number that the child selected, you merely subtract half of 8, that is, 4, from 52 to get 48.

Note that this trick is not dependent on the number 8. It can work nicely with any even number replacing the 8.

How to Determine Your Friend's Birthday

Without telling you the month and day on which he was born, have your friend do the following and he will end up with his birthday. Begin by having your friend multiply the day of his birth by 3, add 5 to that product, and then multiply that sum by 4. Next, add to this product the day plus the month of the birth, and then subtract 20 from that result. Have your friend tell you this result, which you will divide by 13. The quotient that results from this division will be the day of his birth, and the remainder from this division will be the month.

Let's try this with an actual birthdate. We will use October 18 as our hidden birthdate.

Following the rules of the trick indicated above, we have the following calculation:

- Multiply the day by 3 to get $18 \times 3 = 54$.
- Adding 5 to that product gives us $54 + 5 = 59$.
- Then multiply by 4 to get $59 \times 4 = 236$.
- We then add to this number the month and day being sought: $236 + 18 + 10 = 264$.
- Then we subtract 20 from that number: $264 - 20 = 244$.
- When this number is divulged, we divided it by 13 to get $244 \div 13 = 18$ with remainder 10. This indicates that the birthday is the 18th day of the 10th month, or October 18th.

A Trick to Determine Your Friend's Age at the Next Birthday

This trick involves several repeated calculations, albeit simple ones, that will allow you to determine your friend's birthdate and her age on that day. You need to ask your friend to do the following calculations, and when she gives you the end result you will do one simple subtraction and have the required information: birthdate and age on the next birthday.

Here are the instructions to give to your friend.

- Multiply the number of the month by 100.
- To that number add the day and double the result.
- Add 5 to the last number and then multiply by 10.
- Add 23 to the last number and then multiply it by 5.
- Add the friend's age to that result.

- At this point you ask your friend to tell you the final number arrived at, and you subtract 365. You will now have the age at the next birthdate. The last two digits on the right are the age, the next two digits are the date, and the last one or two digits are the month.

Let's try this trick someone who will be 56 years old on the next August 17.

- Multiply the 8th month (August) by 100 to get 800.
- Add 17 and double that result: $800+17=817$; then $817\times 2=1634$.
- Add 5 to that number and multiply by 10: $1634+5=1639$; and then $1639\times 10=16{,}390$.
- Add 23 to that number, then multiply it by 5, and add 56 (your friend's age), that is, $16{,}390+23=16{,}413$; $16413\times 5=82{,}065$; and $82{,}065+56=82{,}121$.
- The number 82,121 has been divulged to you. Now subtract 365 from that number: $82{,}114-365=81{,}756$, which provides you with the necessary information.

We then partition our final result (81,756) as follows: 8/17 and 56, which indicates August 17 age 56. Remember, steps 1 through 4 are done by the friend, and step 5 shows your cleverness by subtracting 365 from your friend's final result to get the required information.

The Trick of Guessing Your Friend's Exact Date of Birth

Now that we have been able to determine age and year of birth, here is a trick to get the exact birthdate. Have your friend follow these instructions:

1. Begin by asking him to write the number of the month in which he was born, but not to divulge it to you.
2. Now have him add the next number in sequence to this number and multiply the sum by 5.

3. Next, you will have him tag on a 0 to the right of the number reached.
4. Ask your friend to select any number less than 100 and tell you what it is. Then he is to add that number to the previous one he calculated.
5. Now ask your friend to add to this new number the day of the month on which he was born.
6. Once again, he is to select any number less than 100, tell you what it is, and tag it on to the right of the result from step 5.
7. To this resulting number, he is to add the number determined by the last two digits of his birth year.

When he tells you the number he has reached, you will be able to tell him the date of his birth. To do that, you need to add 50 to the first number your friend brought into the procedure (step 4). Then place the number he told you in step 6 to the right of the number you just reached. Then take your friend's last number, which he told you, and subtract the number you just developed from it. That will give you your friend's birthdate.

To best understand this process, let's use an illustrative example. Without divulging this to the trickster, we will use the birth date of October 18, 1942.

Without telling the trickster, we begin by writing down the number of our secret birth month, namely, 10. According to the instructions, we will add the next number in sequence and multiply by 5: $(10 + 11) \times 5 = 105$. Next, we are asked to attach a 0 to this number to get 1050. We now tell the trickster that we are adding 36 to our secret number to get 1086. Since our day of birth is the 18th, we add that to our previously obtained number to get $1086 + 18 = 1104$. Once again, we are asked to select a number less than 100, which we will do with the number 28, and tell the trickster what that number is. Then we tag that number onto the right side of the number we previously reached, namely, 1104, so that we get 110,428. We are now asked to add to this number the last two digits of the birth year so that we get $110{,}428 + 42 = 110{,}470$.

Now the trickster begins the calculation: he needs to add 50 to the number that we first added into the process: 50 + 36 = 86. Then he tags on the second number we selected, namely 28, onto this number to get 8628. He subtracts this number from our last number to get 110,470 – 8628 = 101,842. Written in date form he gets 10/18/42, which is October 18, 1942.

Although this process is a bit tedious, it is nonetheless quite an impressive trick. Once again, it will motivate further investigation to see why it works.

How a Telephone Number Can Reveal a Birthyear

To guess someone's birthyear using this trick involves a multistep process. Begin by asking your friend, whose birthyear you will try to discover, to write the last four digits of his telephone number without telling you. Have him scramble these digits and then subtract these two numbers. If the result has more than one digit, then add the digits to get a new number. If this result still has more than one digit, once again, add the digits and continue this process until he reaches a single-digit number. Once a single-digit number has been reached, 8 is to be added to this digit. Then he is to add this number to the last two digits of the birthyear. Subtracting 17 from this result will produce the sought-after birthyear. Therefore, your friend's telephone number has revealed his birthyear, as well as his age!

Let's see how this trick works with the secret birth year of 1948. Remember, we are trying to see how your friend's telephone number can reveal the year 1948. Let's try to apply this to the telephone number 212-567-7879. By scrambling the last four digits we could get the number 9877. Subtracting these two numbers: 9877 – 7879 = 1998.

We now add the digits of this difference:

$$1 + 9 + 9 + 8 = 27.$$

Since we still do not have a single-digit number, we will once again add the digits of this sum: 2 + 7 = 9. To this number we will add 8 to get 9 + 8 = 17. We will now add the last two digits of our birth year: $48 + 17 = 65$. When we subtract 17 from this number 65, we will get 48; so, our birth year is 1948.

This trick is rather easy to explain. When we take any number, rearrange the digits, and then subtract the two resulting numbers, the result will always be a number which is divisible by 9. When a number is divisible by 9, the sum of its digits will have to be divisible by 9 as well. Therefore, adding 17 to the birth year and then subtracting that same amount gets us back to the birth year.

Guessing a Person's Age and Date of Birth

The person whose age and date of birth you're going to determine must do the following:

- Multiply the number of the month of her date of birth by 100.
- Add the date of the month of the birth.
- Multiply the sum by 2 and add 8.
- Multiply this sum by 5 and add 4.
- Multiply the sum by 10 and add 4
- Add her age to this result.
- Subtract 444 from this previous number.

Her age can then be determined, as well as the date of birth.

To further secure understanding of this procedure, consider the following example of a person aged 83 whose birthday is October 18, 1942.

- Multiply the number of the month of her date of birth by 100.

$$10 \times 100 = 1000.$$

- Add the date of the month of the birth.

$$1000 + 18 = 1018.$$

- Multiply the sum by 2 and add 8.

$$(2 \times 1018) + 8 = 2044.$$

- Multiply this sum by 5 and add 4.

$$(5 \times 2044) + 4 = 10{,}224.$$

- Multiply the sum by 10 and add 4.

$$(10{,}224 \times 10) + 4 = 102{,}244.$$

- Add her age to this previous number.

$$102{,}244 + 83 = 102{,}327.$$

- Subtract 444 from this previous number.

$$10{,}307 - 444 = 10\ 18\ \mathbf{83}.$$

The result is clear as it shows the 10th month, the date 18, and the age 83.

Guess Your Age and Address

This trick will impress your friends as you will be able to guess their age and the number of their house address. The trick is conducted by following these simple steps.

Tell your friend to do the following:

- Take the number of the address and double it.
- Add 5 to this number.
- Multiply this number by 50.
- Have the friend add his age to the number.
- Add the number of days in a non-leap year.
- Subtract 615.
- Inspect the result: The two right-most digits represent your friend's age and the remaining numbers represent his address.

To make this clear, we will do an example with an 82-year-old who lives at 634 Maple Street.

- Take the number of the house and double it.

$$634 \times 2 = 1268.$$

- Add 5 to this number.

$$1268 + 5 = 1273.$$

- Multiply this number by 50.

$$1273 \times 50 = 63{,}650.$$

- Have the friend add his age to the number.

$$63{,}650 + 82 = 63{,}732.$$

- Add the number of days in a non-leap year.

$$63{,}732 + 365 = 64{,}097.$$

- Subtract 615.

$$64{,}097 - 615 = 63{,}482.$$

- Inspect the result: the numbers on the right of the result represent your friend's age and the remaining numbers his address. The address is 634 and the age is 82.

Guessing Your Friend's Number Sequence

This number oddity should surely impress your friend with your talent in guessing a number sequence. This oddity involves asking your friend to select any three *consecutive* numbers less than 50. Ask your friend to add these numbers, and to this sum add any multiple of 3. The friend must inform you as to which multiple of 3 was used. Next, have your friend multiply the number obtained by 67. Your friend is then to tell you the last two digits of this result, and you will be able to determine the original sequence with which your friend began.

Now you will show your talent with this oddity. In order to determine the original sequence, you need to take that previously-selected multiple of three which was exposed to you and divide it by 3. Then add 1 to this quotient. Subtract this number from the number determined by the last two digits of your friend's result. The result of this subtraction will yield the first of the three sequential numbers originally selected by the player.

It might be easiest to explain the cleverness of this oddity through an illustrative example. Suppose your friend selects the sequence 7, 8, and 9, and keeps this sequence secret. Your task is to discover the sequence. First, your friend is asked to add the selected numbers, which have a sum of 24. Then he adds a multiple of 3, say 12, to get 36. Next, your friend is told to multiply by this number by 67 to get $36 \times 67 = 2412$. We are now informed that the last two digits are the number 12. Now we divide the previously-selected multiple of 3, which was 12, by 3, giving us 4, to which we then add 1 to get 5. All we need to do now is subtract 5 from the number determined by the last two digits: $12 - 5 = 7$. This result will be the first number of the sequence of three numbers that your friend originally selected. Thus, the sequence your friend selected was 7, 8, 9.

To firm up your understanding of this oddity, we will do one more example. Suppose your friend selects the numbers 13, 14, and 15 as the original sequence. You ask your friend to take the sum of these sequential numbers, which is 42, and then add a multiple of 3, which could be 18, yielding $42 + 18 = 60$. This is then multiplied by 67, giving us 4020. We must remember the last two digits, namely, 20. Then we divide the selected multiple of 3, which was 18, by 3 to get 6. We add 1 to get 7, which we subtract from the last two digits of the result, 20, to get 13, which was the first of the three numbers secretly selected. Therefore we are able to establish that the secret sequence is 13, 14, 15.

Making 100 from the Sequence 1 Through 9

Imagine that you are presented with the following challenge. You have the sequence 123456789, and you must place plus and minus signs between numbers so that the total is 100.

Here are a few examples of how that can be done.

$$123 - 45 - 67 + 89 = 100$$

$$123 + 45 - 67 + 8 - 9 = 100$$

$$123 + 4 - 5 + 67 - 89 = 100$$

$$123 - 4 - 5 - 6 - 7 + 8 - 9 = 100$$

$$12+3+4+5-6-7+89=100$$

$$12+3-4+5+67+8+9=100$$

$$12-3-4+5-6+7+89=100$$

$$1+23-4+5+6+78-9=100$$

$$1+23-4+56+7+8+9=100$$

$$1+2+34-5+67-8+9=100$$

$$1+2+3-4+5+6+78+9=100$$

$$-1+2-3+4+5+6+78+9=100.$$

An Arithmetic Puzzle

Placing math symbols, such as $+$, $-$, $\div$, !, $\sqrt{\ }$, and parentheses, see if you can make these ten lines valid equations.

$$0\quad 0\quad 0\quad =6$$

$$1\quad 1\quad 1\quad =6$$

$$2\quad 2\quad 2\quad =6$$

$$3\quad 3\quad 3\quad =6$$

$$4\quad 4\quad 4\quad =6$$

$$5\quad 5\quad 5\quad =6$$

$$7\quad 7\quad 7\quad =6$$

$$8\quad 8\quad 8\quad =6$$

$$9\quad 9\quad 9\quad =6$$

Here are some possibilities of correct equalities:

$$0!+0!+0!=6$$

$$(1+1+1)!=6$$

$$2+2+2=6$$

$$(3\times 3)-3=6$$

$$\sqrt{4}+\sqrt{4}+\sqrt{4}=6$$

$$(5 \div 5) + 5 = 6$$

$$6 + 6 - 6 = 6$$

$$7 - (7 \div 7) = 6$$

$$\left[\sqrt{(8 \div 8)+8}\right]! = 6$$

$$\left(\sqrt{9}\times\sqrt{9}\right)-\sqrt{9}=6.$$

Conclusion

You have now completed your journey through the oddities of mathematics. These tricks and puzzles are intended to motivate the general populace and educate them about the hidden aspects of mathematics that are unfortunately rarely shared in the school curriculum. It would be good to communicate the experiences encountered in this book with others, so that mathematics gains a popularity which it so clearly merits.

Appendix A

Triangular Numbers

Following is a list of the first 1000 triangular numbers.

1, 3, 6, 10, 15, 21, 28, 36, 45, 55, 66, 78, 91, 105, 120, 136, 153, 171, 190, 210, 231, 253, 276, 300, 325, 351, 378, 406, 435, 465, 496, 528, 561, 595, 630, 666, 703, 741, 780, 820, 861, 903, 946, 990, 1035, 1081, 1128, 1176, 1225, 1275, 1326, 1378, 1431, 1485, 1540, 1596, 1653, 1711, 1770, 1830, 1891, 1953, 2016, 2080, 2145, 2211, 2278, 2346, 2415, 2485, 2556, 2628, 2701, 2775, 2850, 2926, 3003, 3081, 3160, 3240, 3321, 3403, 3486, 3570, 3655, 3741, 3828, 3916, 4005, 4095, 4186, 4278, 4371, 4465, 4560, 4656, 4753, 4851, 4950, 5050, 5151, 5253, 5356, 5460, 5565, 5671, 5778, 5886, 5995, 6105, 6216, 6328, 6441, 6555, 6670, 6786, 6903, 7021, 7140, 7260, 7381, 7503, 7626, 7750, 7875, 8001, 8128, 8256, 8385, 8515, 8646, 8778, 8911, 9045, 9180, 9316, 9453, 9591, 9730, 9870, 10011, 10153, 10296, 10440, 10585, 10731, 10878, 11026, 11175, 11325, 11476, 11628, 11781, 11935, 12090, 12246, 12403, 12561, 12720, 12880, 13041, 13203, 13366, 13530, 13695, 13861, 14028, 14196, 14365, 14535, 14706, 14878, 15051, 15225, 15400, 15576, 15753, 15931, 16110, 16290, 16471, 16653, 16836, 17020, 17205, 17391, 17578, 17766, 17955, 18145, 18336, 18528, 18721, 18915, 19110, 19306, 19503, 19701, 19900, 20100, 20301, 20503, 20706, 20910, 21115, 21321, 21528, 21736, 21945, 22155, 22366, 22578, 22791, 23005, 23220, 23436, 23653, 23871, 24090, 24310, 24531, 24753, 24976, 25200,

25425, 25651, 25878, 26106, 26335, 26565, 26796, 27028, 27261, 27495, 27730, 27966, 28203, 28441, 28680, 28920, 29161, 29403, 29646, 29890, 30135, 30381, 30628, 30876, 31125, 31375, 31626, 31878, 32131, 32385, 32640, 32896, 33153, 33411, 33670, 33930, 34191, 34453, 34716, 34980, 35245, 35511, 35778, 36046, 36315, 36585, 36856, 37128, 37401, 37675, 37950, 38226, 38503, 38781, 39060, 39340, 39621, 39903, 40186, 40470, 40755, 41041, 41328, 41616, 41905, 42195, 42486, 42778, 43071, 43365, 43660, 43956, 44253, 44551, 44850, 45150, 45451, 45753, 46056, 46360, 46665, 46971, 47278, 47586, 47895, 48205, 48516, 48828, 49141, 49455, 49770, 50086, 50403, 50721, 51040, 51360, 51681, 52003, 52326, 52650, 52975, 53301, 53628, 53956, 54285, 54615, 54946, 55278, 55611, 55945, 56280, 56616, 56953, 57291, 57630, 57970, 58311, 58653, 58996, 59340, 59685, 60031, 60378, 60726, 61075, 61425, 61776, 62128, 62481, 62835, 63190, 63546, 63903, 64261, 64620, 64980, 65341, 65703, 66066, 66430, 66795, 67161, 67528, 67896, 68265, 68635, 69006, 69378, 69751, 70125, 70500, 70876, 71253, 71631, 72010, 72390, 72771, 73153, 73536, 73920, 74305, 74691, 75078, 75466, 75855, 76245, 76636, 77028, 77421, 77815, 78210, 78606, 79003, 79401, 79800, 80200, 80601, 81003, 81406, 81810, 82215, 82621, 83028, 83436, 83845, 84255, 84666, 85078, 85491, 85905, 86320, 86736, 87153, 87571, 87990, 88410, 88831, 89253, 89676, 90100, 90525, 90951, 91378, 91806, 92235, 92665, 93096, 93528, 93961, 94395, 94830, 95266, 95703, 96141, 96580, 97020, 97461, 97903, 98346, 98790, 99235, 99681, 100128, 100576, 101025, 101475, 101926, 102378, 102831, 103285, 103740, 104196, 104653, 105111, 105570, 106030, 106491, 106953, 107416, 107880, 108345, 108811, 109278, 109746, 110215, 110685, 111156, 111628, 112101, 112575, 113050, 113526, 114003, 114481, 114960, 115440, 115921, 116403, 116886, 117370, 117855, 118341, 118828, 119316, 119805, 120295, 120786, 121278, 121771, 122265, 122760, 123256, 123753, 124251, 124750, 125250, 125751, 126253, 126756, 127260, 127765, 128271, 128778, 129286, 129795, 130305, 130816, 131328, 131841, 132355, 132870, 133386, 133903, 134421, 134940, 135460, 135981, 136503, 137026, 137550, 138075, 138601, 139128, 139656, 140185, 140715, 141246, 141778, 142311, 142845, 143380, 143916,

144453, 144991, 145530, 146070, 146611, 147153, 147696, 148240, 148785, 149331, 149878, 150426, 150975, 151525, 152076, 152628, 153181, 153735, 154290, 154846, 155403, 155961,156520, 157080, 157641, 158203, 158766, 159330, 159895, 160461, 161028, 161596, 162165, 162735, 163306, 163878, 164451, 165025, 165600, 166176, 166753, 167331, 167910, 168490, 169071, 169653, 170236, 170820, 171405, 171991, 172578, 173166, 173755, 174345, 174936, 175528, 176121, 176715, 177310, 177906, 178503, 179101, 179700, 180300, 180901, 181503, 182106, 182710, 183315, 183921, 184528, 185136, 185745, 186355, 186966, 187578, 188191, 188805, 189420, 190036, 190653, 191271, 191890, 192510, 193131, 193753, 194376, 195000, 195625, 196251, 196878, 197506, 198135, 198765, 199396, 200028, 200661, 201295, 201930, 202566, 203203, 203841, 204480, 205120, 205761, 206403, 207046, 207690, 208335, 208981, 209628, 210276, 210925, 211575, 212226, 212878, 213531, 214185, 214840, 215496, 216153, 216811, 217470, 218130, 218791, 219453, 220116, 220780, 221445, 222111, 222778, 223446, 224115, 224785, 225456, 226128, 226801, 227475, 228150, 228826, 229503, 230181, 230860, 231540, 232221, 232903, 233586, 234270, 234955, 235641, 236328, 237016, 237705, 238395, 239086, 239778, 240471, 241165, 241860, 242556, 243253, 243951, 244650, 245350, 246051, 246753, 247456, 248160, 248865, 249571, 250278, 250986, 251695, 252405, 253116, 253828, 254541, 255255, 255970, 256686, 257403, 258121, 258840, 259560, 260281, 261003, 261726, 262450, 263175, 263901, 264628, 265356, 266085, 266815, 267546, 268278, 269011, 269745, 270480, 271216, 271953, 272691, 273430, 274170, 274911, 275653, 276396, 277140, 277885, 278631, 279378, 280126, 280875, 281625, 282376, 283128, 283881, 284635, 285390, 286146, 286903, 287661, 288420, 289180, 289941, 290703, 291466, 292230, 292995, 293761, 294528, 295296, 296065, 296835, 297606, 298378, 299151, 299925, 300700, 301476, 302253, 303031, 303810, 304590, 305371, 306153, 306936, 307720, 308505, 309291, 310078, 310866, 311655, 312445, 313236, 314028, 314821, 315615, 316410, 317206, 318003, 318801, 319600, 320400, 321201, 322003, 322806, 323610, 324415, 325221, 326028, 326836, 327645, 328455, 329266, 330078, 330891, 331705, 332520, 333336, 334153, 334971, 335790, 336610, 337431, 338253, 339076, 339900,

340725, 341551, 342378, 343206, 344035, 344865, 345696, 346528, 347361, 348195, 349030, 349866, 350703, 351541, 352380, 353220, 354061, 354903, 355746, 356590, 357435, 358281, 359128, 359976, 360825, 361675, 362526, 363378, 364231, 365085, 365940, 366796, 367653, 368511, 369370, 370230, 371091, 371953, 372816, 373680, 374545, 375411, 376278, 377146, 378015, 378885, 379756, 380628, 381501, 382375, 383250, 384126, 385003, 385881, 386760, 387640, 388521, 389403, 390286, 391170, 392055, 392941, 393828, 394716, 395605, 396495, 397386, 398278, 399171, 400065, 400960, 401856, 402753, 403651, 404550, 405450, 406351, 407253, 408156, 409060, 409965, 410871, 411778, 412686, 413595, 414505, 415416, 416328, 417241, 418155, 419070, 419986, 420903, 421821, 422740, 423660, 424581, 425503, 426426, 427350, 428275, 429201, 430128, 431056, 431985, 432915, 433846, 434778, 435711, 436645, 437580, 438516, 439453, 440391, 441330, 442270, 443211, 444153, 445096, 446040, 446985, 447931, 448878, 449826, 450775, 451725, 452676, 453628, 454581, 455535, 456490, 457446, 458403, 459361, 460320, 461280, 462241, 463203, 464166, 465130, 466095, 467061, 468028, 468996, 469965, 470935, 471906, 472878, 473851, 474825, 475800, 476776, 477753, 478731, 479710, 480690, 481671, 482653, 483636, 484620, 485605, 486591, 487578, 488566, 489555, 490545, 491536, 492528, 493521, 494515, 495510, 496506, 497503, 498501, 499500, 500500,

Appendix B

Pythagorean Triples

(3, 4, 5)	(5, 12, 13)	(7, 24, 25)	(8, 15, 17)	(9, 40, 41)
(11, 60, 61)	(12, 35, 37)	(13, 84, 85)	(15, 112, 113)	(16, 63, 65)
(17, 144, 145)	(19, 180, 181)	(20, 21, 29)	(20, 99, 101)	(21, 220, 221)
(23, 264, 265)	(24, 143, 145)	(25, 312, 313)	(27, 364, 365)	(28, 45, 53)
(28, 195, 197)	(29, 420, 421)	(31, 480, 481)	(32, 255, 257)	(33, 56, 65)
(33, 544, 545)	(35, 612, 613)	(36, 77, 85)	(36, 323, 325)	(37, 684, 685)
(39, 80, 89)	(39, 760, 761)	(40, 399, 401)	(41, 840, 841)	(43, 924, 925)
(44, 117, 125)	(44, 483, 485)	(48, 55, 73)	(48, 575, 577)	(51, 140, 149)
(52, 165, 173)	(52, 675, 677)	(56, 783, 785)	(57, 176, 185)	(60, 91, 109)
(60, 221, 229)	(60, 899, 901)	(65, 72, 97)	(68, 285, 293)	(69, 260, 269)
(75, 308, 317)	(76, 357, 365)	(84, 187, 205)	(84, 437, 445)	(85, 132, 157)
(87, 416, 425)	(88, 105, 137)	(92, 525, 533)	(93, 476, 485)	(95, 168, 193)
(96, 247, 265)	(100, 621, 629)	(104, 153, 185)	(105, 208, 233)	(105, 608, 617)
(108, 725, 733)	(111, 680, 689)	(115, 252, 277)	(116, 837, 845)	(119, 120, 169)
(120, 209, 241)	(120, 391, 409)	(123, 836, 845)	(124, 957, 965)	(129, 920, 929)
(132, 475, 493)	(133, 156, 205)	(135, 352, 377)	(136, 273, 305)	(140, 171, 221)
(145, 408, 433)	(152, 345, 377	(155, 468, 493)	(156, 667, 685)	(160, 231, 281)
(161, 240, 289)	(165, 532, 557)	(168, 425, 457)	(168, 775, 793)	(175, 288, 337)
(180, 299, 349)	(184, 513, 545)	(185, 672, 697)	(189, 340, 389)	(195, 748, 773)
(200, 609, 641)	(203, 396, 445)	(204, 253, 325)	(205, 828, 853)	(207, 224, 305)
(215, 912, 937)	(216, 713, 745)	(217, 456, 505)	(220, 459, 509)	(225, 272, 353)

Appendix C
Prime Numbers

2	3	5	7	11	13	17	19	23	29
31	37	41	43	47	53	59	61	67	71
73	79	83	89	97	101	103	107	109	113
127	131	137	139	149	151	157	163	167	173
179	181	191	193	197	199	211	223	227	229
233	239	241	251	257	263	269	271	277	281
283	293	307	311	313	317	331	337	347	349
353	359	367	373	379	383	389	397	401	409
419	421	431	433	439	443	449	457	461	463
467	479	487	491	499	503	509	521	523	541
547	557	563	569	571	577	587	593	599	601
607	613	617	619	631	641	643	647	653	659
661	673	677	683	691	701	709	719	727	733
739	743	751	757	761	769	773	787	797	809
811	821	823	827	829	839	853	857	859	863
877	881	883	887	907	911	919	929	937	941
947	953	967	971	977	983	991	997	1009	1013
1019	1021	1031	1033	1039	1049	1051	1061	1063	1069
1087	1091	1093	1097	1103	1109	1117	1123	1129	1151
1153	1163	1171	1181	1187	1193	1201	1213	1217	1223

(*Continued*)

(*Continued*)

1229	1231	1237	1249	1259	1277	1279	1283	1289	1291
1297	1301	1303	1307	1319	1321	1327	1361	1367	1373
1381	1399	1409	1423	1427	1429	1433	1439	1447	1451
1453	1459	1471	1481	1483	1487	1489	1493	1499	1511
1523	1531	1543	1549	1553	1559	1567	1571	1579	1583
1597	1601	1607	1609	1613	1619	1621	1627	1637	1657
1663	1667	1669	1693	1697	1699	1709	1721	1723	1733
1741	1747	1753	1759	1777	1783	1787	1789	1801	1811
1823	1831	1847	1867	1861	1871	1873	1877	1879	1889
1901	1907	1913	1931	1933	1949	1951	1973	1979	1987
1993	1997	1999	2003	2011	2017	2027	2029	2039	2053
2063	2069	2081	2083	2087	2089	2099	2111	2113	2129
2131	2137	2141	2143	2153	2161	2179	2203	2207	2213
2221	2237	2239	2243	2251	2267	2269	2273	2281	2287
2293	2297	2309	2311	2333	2339	2341	2347	2351	2357
2371	2377	2381	2383	2389	2393	2399	2411	2417	2423
2437	2441	2447	2459	2467	2473	2477	2503	2521	2531
2539	2543	2549	2551	2557	2579	2591	2593	2609	2617

Index

www.ingramcontent.com/pod-product-compliance
Lightning Source LLC
LaVergne TN
LVHW050927210926
848482LV00013B/562
* 9 7 8 9 8 1 9 8 2 3 1 8 5 *